Teaching and Studying the Americas

Teaching and Studying the Americas

Cultural Influences from Colonialism to the Present

Edited by
*Anthony B. Pinn, Caroline F. Levander,
and Michael O. Emerson*

TEACHING AND STUDYING THE AMERICAS
Copyright © Anthony B. Pinn, Caroline F. Levander, and Michael O. Emerson 2010.

All rights reserved.

First published in 2010 by
PALGRAVE MACMILLAN®
in the United States—a division of St. Martin's Press LLC,
175 Fifth Avenue, New York, NY 10010.

Where this book is distributed in the UK, Europe and the rest of the world, this is by Palgrave Macmillan, a division of Macmillan Publishers Limited, registered in England, company number 785998, of Houndmills, Basingstoke, Hampshire RG21 6XS.

Palgrave Macmillan is the global academic imprint of the above companies and has companies and representatives throughout the world.

Palgrave® and Macmillan® are registered trademarks in the United States, the United Kingdom, Europe and other countries.

ISBN: 978–0–230–61512–0

Library of Congress Cataloging-in-Publication Data

    Teaching and studying the Americas : cultural influences from colonialism to the present / edited by Anthony B. Pinn, Caroline F. Levander, and Michael O. Emerson.
      p. cm.
    ISBN 978–0–230–61512–0 (hardback)
    1. America—Study and teaching (Higher) 2. Western Hemisphere—Study and teaching (Higher) I. Pinn, Anthony B. II. Levander, Caroline Field, 1964– III. Emerson, Michael O., 1965–

E16.5.T43 2010
918.1′2—dc22   2010016613

A catalogue record of the book is available from the British Library.

Design by Newgen Imaging Systems (P) Ltd., Chennai, India.

First edition: November 2010

10 9 8 7 6 5 4 3 2 1

Printed in the United States of America.

Contents

Acknowledgments vii

Introduction 1
*Caroline F. Levander, Anthony B. Pinn,
and Michael O. Emerson*

Part 1   Locating and Dislocating the Americas

1 Good Neighbor/Bad Neighbor: Boltonian Americanism
   and Hemispheric Studies 13
   Antonio Barrenechea

2 Bad Neighbor/Good Neighbor: Across the Disciplines
   toward Hemispheric Studies 27
   Caroline F. Levander

3 Coloniality at Large: The Western Hemisphere and
   the Colonial Horizon of Modernity 49
   Walter D. Mignolo (Translated by Michael Ennis)

Part 2   Disciplining Hemispheric Studies

4 A Major Motion Picture: Studying and Teaching
   the Americas 77
   Michael O. Emerson

5 Embodied Meaning: The "Look" and "Location" of
   Religion in the American Hemisphere 93
   Anthony B. Pinn

6 Primeval Whiteness: White Supremacists, (Latin)
   American History, and the Trans-American Challenge to
   Critical Race Studies 109
   Ruth Hill

7 The Making of "Americans": Old Boundaries,
   New Realities   139
   Karen Manges Douglas and Rogelio Saenz

8 Interdisciplinary Approaches to Teaching the History
   of the Western Hemisphere   157
   Moramay López-Alonso

Part 3   Programs and Pedagogy

9 Beyond National Borders: Researching and Teaching
   Jovita González   179
   Heather Miner and Robin Sager

10 Migrant Archives: New Routes in and out of
   American Studies   199
   Rodrigo Lazo

11 Partnering Across the Americas: Crossing National and
   Disciplinary Borders in Archival Development   219
   Melissa Bailar

12 Ghosts of the American Century: The Intellectual,
   Programmatic, and Institutional Challenges for
   Transnational/Hemispheric American Studies   243
   Deborah Cohn and Matthew Pratt Guterl

Works Cited   263

Author Biographies   283

Index   287

Acknowledgments

The editors cannot list everyone who assisted in some way with this book. However, we extend our gratitude to each and every one of them, particularly our editors at Palgrave Macmillan. Their patience is greatly appreciated. We must also thank two Rice University graduate students, AnaMaria Seglie and Darrius Hills, for their assistance with preparing this volume for publication. Caroline Levander would like to acknowledge the Huntington Library for invaluable resources for the primary material upon which she draws in her chapter. In addition, Melissa Bailar would like to thank Geneva Henry and Monica Rivero for supplying high resolution versions of several images used in her essay and securing their availability under the Creative Commons license.

The editors would like to thank the contributors to this volume for their participation and keen insights. We must also extend our deep gratitude for the assistance received from Alexander Byrd, our colleague and friend, for the many hours he put into helping us with arranging, preparing, and reviewing the project. This would have been a much more difficult and far less pleasant project without his assistance. Thank you.

Finally, the editors acknowledge permission to reprint the following pieces (in order of appearance):

Antonio Barrenechea, "Good Neighbor/Bad Neighbor: Boltonian Americanism and Hemispheric Studies," in *Comparative Literature*, Volume 61, Number 3 (2009): 231–243. Used by permission of the publisher, Duke University Press.

Walter D. Mignolo, "Coloniality at Large: The Western Hemisphere in the Colonial Horizon of Modernity." This work originally appeared in *The New Centennial Review*, Volume 1, Number 2 (Fall 2001): 19–54. The journal is published by Michigan State University Press and the article is included here by permission of the MSU Press.

Rodrigo Lazo, "Migrant Archives," 36–54. From *States of Emergency: The Object of American Studies*, edited by Russ Castronovo and Susan Gillman. Copyright © 2009 by the University of North Carolina Press. Used by permission of the publisher.

Introduction

Caroline F. Levander, Anthony B. Pinn, and Michael O. Emerson

This book grows out of intellectual debate, inquiry, and struggle in which a group of Rice University faculty engaged over the course of six years. Our collaborations focused on the significance of the American hemisphere for teaching and research. Starting with conversations and the sharing of pedagogical strategies, this concern with the implications of hemispheric studies resulted in the development of the Rice University Americas Colloquium.[1] But if our collaboration started at one institution, we were immediately aware of how many other universities and faculty across North America were engaged in similar thought experiments. We have collected many of their stories and voices within this volume.

Context for This Project

As we learned more about the institutional and geographic contexts in which colleagues at other universities were undertaking similar endeavors, we came to see how unique each set of answers to these shared questions must be. The relative small size of Rice and the relative ease with which faculty were able to move between disciplines raised a compelling set of questions and possibilities for the colloquium. Recognizing the degree to which studies in all academic disciplines have increasingly begun to break down the traditional boundaries of studies of the different regions of the Americas, it was clear that attention to the inter-Americas could be a source of intellectual synergy and distinction. We believed (and continue to believe) that Rice is well situated to offer dynamic work in this area, both because of its physical geography (located in the "gateway" and deeply cosmopolitan city of Houston) and intellectual geography (faculty in history, literature, Southern studies, religion, and urban sociology).

Through on- and off-campus collaborations, guest speakers, curriculum development, and technological innovation, the Americas Colloquium pushed for increasingly creative approaches to the study and teaching of the Americas. Most broadly, the Americas Colloquium considered how interdisciplinary conversation, critique, and collaboration might enrich and transform humanities graduate education among Americanists. The colloquium approached the present and future practice of American cultural analysis in a postnational and global studies context; hence, the colloquium adopted an approach that considered the parameters and futures of American cultural practice from a number of different disciplinary vantage points. This work pointed out in rather clear ways the manner in which interdisciplinary and hemispheric work might enrich and transform both research and teaching.

Moving beyond the confines of Rice University and the Americas Colloquium, this volume is an effort to explore and interrogate this important question. So conceived, this is a volume about expanding scholarly vision and practice. Over the past two decades, important work in academic disciplines increasingly has been breaking down many of the regional and national boundaries that have long characterized the study of the Americas. Future research and curriculum on the hemisphere will, no doubt, continue to emphasize comparative and cross-regional studies. New graduate and undergraduate programs at institutions such as University of Southern California, Indiana University, and the University of Toronto; new journals, such as *The Americas: A Quarterly Review of Inter-American Cultural History, Comparative American Studies* and *Review of International American Studies*; and new associations, such as the International American Studies Association, mark a dramatic shift in focus away from nation-based frameworks. Interdisciplinary work that moves beyond analysis of any one nation in isolation and that places urgent intellectual questions in the larger matrix of the Americas as a hemisphere has begun to assume academic prominence across humanities and social science disciplines. This textured, comparative study has promoted creative, original insights into the full complexities of life and thought in the Americas.

While outstanding universities and scholars are producing innovative and paradigm-shifting research, and new journals provide critical venues for article-length scholarship that captures this shift in intellectual perspective, there remains limited attention to the overarching methodological, institutional, and pedagogical issues resulting from the growth of inter-American, or American hemispheric studies. Current developments in scholarship wrestle with the nature and meaning, the content and theoretical arrangements of the Americas, marking the tension between nation and hemisphere. And courses are developed within new programs that highlight hemispheric study. What remains undone, however, is interrogation of central questions that these new developments raise for universities, students, and teachers: How does attention to the American hemisphere as an overarching conceptual framework, and the comparative and connective analysis such a framework encourages, affect issues of institution building, research method, academic practice, and pedagogy? How does one teach or research the Americas? How do courses with traditional U.S. foci (the U.S. Literature Survey and the U.S. History Survey, for example) engage other, often lost or marginalized stories? What different methods of analysis are needed when these stories become part of our teaching toolbox? How do research databases address the challenges of multilingualism that an Americas approach raises? Is it possible to develop research tools and to teach the more complex, multilayered, and often obscured literary, religious, and social histories of the Americas, given existing institutional and curricular constraints?

Certainly individual courses and research initiatives are being developed, but to date we have precious little extended consideration of the full range of fundamental challenges and questions that a hemispheric studies perspective raises for university teachers, scholars, students, and administrators. While recent books such as Caroline Levander and Robert Levine's *Hemispheric American Studies*, Winfried Siemerling and Sarah Phillips Casteel's *Canada and Its Americas: Transnational Navigations*, and Sandhya Shukla and Heidi Tinsman's *Imagining Our Americas: Toward a Transnational Frame* explore disciplinary shifts towards a hemispheric or Americas frame of reference, this collection is unique in its primary concern with questions of institutional practice, pedagogic transformation, and research perspectives. To facilitate this focus, our contributors not only represent a diverse set of disciplines (religious studies, history, English, sociology, Spanish and Portuguese, American studies, African American and African Diaspora studies), but they also bring a diverse range of institutional perspectives and expertise (for example, professors, graduate students, and archivists) to the complex questions with which the collection concerns itself.

This volume is organized into three sections. The first section—Locating and Dislocating the Americas—provides a broad overview of the conversations surrounding the idea of the Western Hemisphere, hemispheric studies, and disciplinary innovation over the course of the last few decades. These overview essays serve to frame the current Americas debates, theoretical perspectives, and scholarly stakes. The reader will come to see from these essays that the theoretical perspectives are not yet set. Instead, the essays bring the reader into the conversation and introduce the current dialogue. Antonio Barrenechea's essay provides a long view of the evolution of a hemispheric perspective, beginning with Herbert Bolton and extending to border studies and current critical attention to hemispheric studies. Caroline Levander explores the tensions as well as possibilities implicit in this move for those operating in the cognate fields comprising the humanities and humanistically inclined social sciences. Her essay contemplates when and how it makes sense to adopt hemispheric frames of reference and the stakes of doing so at the current scholarly moment. Finally, Walter Mignolo's essay reminds us of the broad sweeps of modernity against which the Western Hemisphere emerges as legible form. Collectively, the essays comprising this section are intended to showcase the conversation as it currently stands, and to do so by having essays authored by scholars from the different perspectives and disciplines of comparative literature, American studies, and Latin American studies.

In the second section—Disciplining Hemispheric Studies—we see that once we think in terms of the Americas, individual disciplines look somewhat different. The essays comprising this section emphasize different aspects of disciplinarity, and they ask different questions. This section provides what may be seen as a series of case studies of how scholars housed in departments of history, religious studies, Spanish/Portuguese, and sociology approach the field. And the second section offers a practicum in disciplinary innovation that can result when one works within the Americas framework. Whether it is motion or the structure of laws for sociology, the body for religious meaning, disease/public health for history, or white supremacy writ hemispheric large for language studies scholars, the Americas perspective offers fruit for scholarly advance. For example, in the Douglas and Saenz essay, the authors trace the ways in which two forms of policies—immigration laws and racial classifications—have shaped who from the Americas is allowed into the United States and how immigrants are subsequently treated. These policies have profound impact for wealth and inequality throughout the Americas.

The final section moves from theory to practice, contemplating different aspects of implementation of an Americas intellectual agenda. Whether it is archive, new course development, or the creation of a hemispheric studies curriculum, the essays in the final section focus on the pragmatic realities as well as the deep intellectual commitments that make teaching and studying the Americas more than a theoretical interest for a wide range of academia's practitioners.

This volume is centrally interested both in the large conceptual opportunities offered by teaching and studying the Americas, and in the nuts-and-bolts realities of actually implementing this perspective—of realizing an Americas model in the institutional infrastructure that universities build (such as library archives, course curricula, and major requirements). Two of the essays in the collection, for example, attend to the very real complexities of collecting, collating, disseminating, and storing Americas' knowledge. Both Melissa Bailar and Rodrigo Lazo begin with an analysis of the significant challenges posed by long-standing institutional practices of using nation-based categories for organizing and disseminating extant knowledge. But they then offer exciting alternatives for reconceptualizing our collective shared knowledge base through the generation of new modes of organizing and representing archival material. Both also recognize the challenge of multilingualism to the collection and dissemination of archival material, and offer some preliminary thoughts on how to integrate multiple languages in the English-speaking framework in which U.S. universities operate. Among other important contributions, Matthew Guterl and Deborah Cohn offer a useful perspective on this complex question of multilingual education in their essay on creating and implementing a hemispheric studies curriculum at a U.S. institution of higher education that has been long known for its nation-based American Studies program.

Not only does teaching and studying the Americas challenge us to grow new research tools and generate new kinds of curricula, but it also brings disciplinary practitioners into new interdisciplinary collaborations and encourages new questions within individual disciplines. To a large extent, the move in American studies toward a hemispheric or Americas model has been championed by those in English and, to a somewhat lesser extent, those in History departments. Consideration, for example, of the Latin American origins and legacies in the development of the United States has led historians and literary studies scholars to think about how local spaces within what is now the United States are tied to other parts of the hemisphere, such that we cannot understand the cultural contours and historical development of a place like Houston, Texas, for example, without understanding the forces at work in the shaping of a place like the Dominican Republic. Isolated points within the United States are no longer completely legible within or encompassed by the nation that currently seems to contain them, but are rather understood as points along a more extensive geographic constellation of peoples, ideas, cultures, and traditions. The essay cowritten by Heather Miner and Robin Sager suggests the rich possibility of this kind of approach for the study of such seemingly separate disciplines as history and English. By reclaiming the literary and historical legacy of exactly the kind of figure that Lazo and Bailar seek to make visible in the official archival record, Miner and Sager not only suggest the utility of a little-known Americas figure for both history and literature course curriculum, but, just as importantly, they model the kind of interdisciplinary collaboration that a commitment to teaching and studying the Americas, of necessity, stimulates.

While Sager and Miner represent the new kinds of interdisciplinary collaborations that can and do emerge when we adopt an Americas approach within disparate disciplines, several pieces in this volume also suggest how attention to teaching and studying the Americas can generate significant methodological shifts of the disciplinary lens within such disciplines as religious studies and sociology. For example, in these disciplines, comparative work is increasingly common, but the focus tends to remain on chipping away at U.S. dominance, or tends to float above questions of geography. By this we mean that these two disciplines often work in a way that seeks to think through and project modalities of liberation and freedom but without sustained attention to the manner in which physical and symbolic geographies shape and influence oppression and schemes of liberation or freedom within, between, and over against the nation/state as the primary modality of identity and meaning. What such work does not do is raise more fundamental questions about the notion of nation or about how one can conceptualize and use an Americas perspective. Essays such as those by Anthony Pinn and Michael Emerson suggest that new approaches, questions, and methods can emerge when sociologists and religious studies scholars operate within an Americas frame and, therefore, that much of value can be learned by thinking theoretically about the Americas as a whole, rather than by approaching the Americas as separable pieces that we then compare. In these essays, what emerges are two new concepts for studying the Americas. Pinn suggests that a focus on the body can serve as the conceptual tool by which to understand the meaning and practice of religion across the Americas. Emerson suggests that utilizing the conceptual tool of motion allows scholars to engage in the study of the political, cultural, and social aspects of the Americas in a cohesive fashion, and also provides a framework for teaching the Americas. Collectively they suggest that new methods, questions, and approaches can and do emerge within individual fields as the extant field imaginary shifts toward hemispheric rather than nation-based frames of reference.

Just as hemispheric studies can alter the questions that disciplines ask and the frames of reference within which those questions are conceptualized, assessed, and answered, so, too, do disciplines exist in dynamic relation to geopolitical terrains, continuing to shape and innovate how we understand and approach the Americas. No better instance of this can be found in the collection than in the work of Moramay López-Alonso, whose essay, among other things, highlights the rich interplay between disciplinary histories and geopolitical understandings of the Western Hemisphere. Through analysis of the Americas' demographic transformation after the arrival of Europeans, López-Alonso shows how including findings from disciplines outside of traditional history—such as public health and climatology—can generate more comprehensive understandings of "historical" events, like the onset of disease and epidemics resulting from the mutation of indigenous germs as well as environmental alterations produced by human factors. Rather than the historical fact of biological disease, Ruth Hill takes up the concept of racial hierarchy and, more particularly, white supremacy, showing how it emerges as a fully blown socioscientific concept within a hemispheric rather than national historical context. Hill illustrates how malleable white supremacy becomes in a trans-American frame, cutting multiple ways and absorbing those of Spanish and Portuguese as well as of Asian ancestry into a multiform and multidisciplinary logic that draws from scientific finding as well as historical precedent in order to naturalize its claims. As the essays of López-Alonso and Hill remind us, disciplines exist in dynamic tension with geopolitical formations and an Americas frame can, therefore, provide strategic opportunities to reconfigure what we think we know about both our disciplinary fields of inquiry and the geographic fields in which we conceptualize problems within individual disciplines.

Purpose: We are mindful that we are asking these questions about teaching and studying the Americas within the particular geoinstitutional context of U.S. universities and academia. Our unavoidable frame of reference for conceptualizing how to approach Americas research and pedagogy is that of the U.S. university, replete with its expectations for undergraduate training and career preparation, faculty promotion and tenure, and graduate student professionalization. U.S.

universities are the place from which we ask these challenging questions, and they are the places where we test, implement, and play out the answers to these questions. We purposefully delimited our field of inquiry to U.S. academia because its protocols are those we know best and because, as many essays point out, funding structures, professional metrics, institutional assessment, and job placement of students remain largely nation-based even as the subject and content of fields of study within the university do not. Yet the tension between our subject and our laboratory is one that need not be constant. Many universities claim to be increasingly global in their aspiration, and if the global university is a concept quickly gaining ground among administrators, it behooves faculty to shape these institutional efforts in ways that further rather than compromise the intellectual authenticity and integrity of scholarly inquiry.

Many United Kingdom and other European institutions, for example, might be understood to be ahead of U.S. institutions in this regard. Consider Oxford University's Rothermere American Institute, the University of Leipzig's Institute for American Studies, or Freie Universität Berlin's John F. Kennedy Institute for North American Studies. All, in different ways, have long placed their curricular focus on the United States in various kinds of transnational and global contexts. An exciting range of new consortia, such as New York University's Hemispheric Institute of Performance and Politics, bring artists, students, and activists throughout the Americas together in various projects. The Hemispheric Institute of Performance and Politics's express purpose is to "create spaces and opportunities for cross-cultural collaboration and interdisciplinary innovation among researchers and practitioners interested in the relationship between performance, politics, and social life in the hemisphere."[2] Likewise, the Tepoztlán Institute for the Transnational History of the Americas hosts faculty and graduate students for weeklong, in-depth discussions of theoretical and historical work related to transnational frameworks. Both institutes are ambitious and laudable initiatives. But neither are organizations primarily committed to pedagogy, rendering them less directly pertinent for our purposes. We recognize, though, that it is precisely these kinds of initiatives that will facilitate the transformation of pedagogic practice at every level of education.

Most generally, this book considers how interdisciplinary conversation, critique, and collaboration enrich and transform humanities and social science education for those teaching and studying traditional Americanist fields. The transition from a national to a hemispheric

American studies is both exciting and daunting. On the one hand, it promises to reinvigorate existing fields. On the other, it poses a serious challenge to received models of intellectual training, research, evaluation, and curricular development. Although many now recognize the importance of this transformation, there is scant institutional support for those teachers who want to reconceive their intellectual work within the rubric of this new research area. As a result, faculty interested in reinventing existing coursework and developing new research models do so without the benefit of senior mentors. Our book fills an essential need by providing a forum for thinking through the implications of this shift for scholars who are at the beginning of their careers and may lack mentorship and institutional support at their home institutions.

The essays in *Teaching and Studying the Americas* introduce students and faculty working in traditional disciplines to the new possibilities for Americanist study opened up when "America" is understood not as a synonym for an isolated United States but as a network of cultural influences that have extended across the hemisphere from the period of colonization to the present. Collectively, the essays comprising *Teaching and Studying the Americas* explore the points of contact between and across the individual disciplines of history, religion, sociology, and literary studies as well as between and across disparate parts of the hemisphere. The essays are more than the sum of their parts because they explore larger questions of disciplinary as well as national boundaries within an Americas paradigm. These authors contemplate the disciplinary boundaries that can at times separate fields of knowledge, but they also explore the evolving methods that emerge once national boundaries are understood as overlapping and multiform.

Notes

1. See the Americas Colloquium website: http://americascolloquium.rice.edu/.
2. Taken from the Institute's webpage at http://hemisphericinstitute.org/hemi/en/mission, accessed January 5, 2010.

PART 1

Locating and Dislocating the Americas

CHAPTER 1

Good Neighbor/Bad Neighbor: Boltonian Americanism and Hemispheric Studies

Antonio Barrenechea

In "The Epic of Greater America," his 1932 presidential address to the American Historical Association, Herbert Eugene Bolton, a historian from Berkeley, proposed that American history should be approached not as a national narrative, but as a story of hemispheric proportions. Bolton's speech was a response to *The Epic of America*, a bestselling historical monograph written by James Truslow Adams in 1931. Adams was a leading historian of the period and had received a Pulitzer Prize in 1922 for his work on settlement, revolution, and nation-building in New England. Although in *The Epic of America*, the study in which he coined the phrase "The American Dream," Adams had extended his investigations beyond the Puritan tradition, his 1931 study remained focused on the United States. By contrast, Bolton's "The Epic of Greater America" transcended the teleology of a single nation to include the interregional and international development of the entire Western Hemisphere.[1]

The main presupposition of "The Epic of Greater America" is that historians mistake the nation for the hemisphere when they equate the United States with America. According to Bolton, a syncretic approach that provides "a broader treatment of American history...to supplement the purely nationalistic presentation to which we are accustomed" is not just ethical or politically advantageous, but "desirable from the standpoint of correct historiography" (448). Without the inter-American perspective, historians can only form a fractured and distorted picture of the United States: "[E]ach local story will have much clearer meaning when studied in the light of the others; [for] much of what has been written of each national history is but a thread out of a larger strand" (449). Guided by the nomenclature, geography, and European lineage of the hemisphere as a whole, Bolton focuses most intently on the colonial period in order to show how nation-centered narratives depended on the erasure of the multiethnic underpinnings of—and in particular the Spanish contributions to—what historians called "America."

Thus, Bolton stresses in "The Epic of Greater America" common patterns of exploration, colonial experience, wars of independence, and nation building. He highlights the interdependent and overlapping histories of Europe and America from the early modern period, when the Spanish established their presence in the New World in ways that altered the U.S. national narrative. With respect to the westward advance, for instance, only after discussing the sixteenth-century Spanish exploration of lands that would later become part of the United States does Bolton extend his argument into the national narrative:

> Most of our American explorer heroes of the Far West, from Smith to Frémont, were in reality belated explorers of a foreign country. For a quarter century after 1820 these trespassers roamed the western wilds, profiting by the fur trade, and "discovering" the mountain passes—which Spaniards had discovered long before. (465)

Unfortunately, the quotation marks around "discovering" do not indicate that Bolton was aware of his Eurocentric view of exploration, nor do they signal U.S. efforts to overtake Native American communities that had already "discovered" their own lands. They do, however, indicate the belatedness of the British arrival in America. Bolton establishes Spanish colonization as the starting point for U.S. history as opposed to the arrival of British separatists in the northeast.

Bolton highlights comparable national and regional narratives through a synchronic approach that observes multiple points of origin. Referencing U.S. independence, he claims that the Spanish founding of the viceroyalty of La Plata, with Buenos Aires as its capital city, was also (and equally) "one of the significant American events of 1776," and "did much to determine the destiny of the southern continent" (454). He reminds historians that "[a] few days before the Declaration of Independence was proclaimed in Philadelphia, San Francisco was founded [by Spain] on the Pacific Coast" and that this "was another of the significant events of 1776" (457). Far from recounting the typical exceptionalist history of Anglo-American expansion, "The Epic of Greater America" features multiple European agents caught up in parallel and intertwining imperial projects. Instead of interpreting British American independence as an event that begins a singular national teleology, Bolton sees it as the first successful revolution among many in the Americas—as a catalyst, in short, for a bloody chain of events that "did not end in Yorktown" (458). He conceives of revolution, then, as a hemispheric matter, as an articulated movement that, however precariously, ousts Europe from America and brings the colonial period to a close. In "Defensive Spanish Expansion and the Significance of the Borderlands," an address delivered three years earlier (1929), Bolton had made the same point by connecting historical figures from the United States to various parts of Latin America and erasing the privileged status of the former: "Washington and his associates merely started the American Revolution; Miranda, Bolívar, San Martín, Hidalgo, Morelos, and Iturbide carried it through" (*Wider Horizons* 67).

Bolton's account of "greater America" was derived from his pioneering work on the Iberian presence along the southern and western fringes of the United States, which he called the "Spanish borderlands." His 1921 book *The Spanish Borderlands: A Chronicle of Old Florida and the Southwest* amplified the famous thesis pronounced in 1893 by Frederick Jackson Turner, one of Bolton's mentors at the University of Wisconsin. In his own address to the American Historical Association, "The Significance of the Frontier in American History," Turner had famously asserted that the experience of settlement and expansion from Europe to the Pacific bolstered the development of a distinct national character. Bolton challenged and revised Turner's argument to include a pantheon of Spanish pioneers—Coronado and Ponce de León, for example—thus anticipating (and in some sense founding) the field of Chicano and border studies.

While conducting archival research in Mexico and the United States, Bolton realized that U.S. historians had perpetuated a "black legend" of Iberian backwardness and Catholic fanaticism in which the Spanish figured as gold-seeking caricatures who lacked the organizational principles necessary to form viable societies in the New World. According to Bolton, these accounts attested to a myopic nationalism among U.S. academicians and were

> the inevitable result of writing United States history in isolation, apart from its setting in the history of the entire Western Hemisphere, of which the United States are but a part. It was the logical corollary of restricting the study of American history to the region between the

forty-ninth parallel and the Gulf of Mexico, as though that area were an inclusive and exclusive entity, and were synonymous with America. (*Wider Horizons* 56)

Assuming a broader perspective, Bolton argued that the Spanish used the presidio and the mission to establish a long-lasting presence in North America. The presidio created a military front against Native Americans and European powers, while the mission represented more than a proselytizing effort; it was a colonial apparatus that helped administer an overseas empire. Representing what Bolton calls "a conspicuous feature of Spain's frontiering genius" (*Wider Horizons* 148), the missions and monasteries created by Jesuits, Franciscans, Dominicans, and other religious orders served administrative functions that included regulating Amerindian societies and chronicling expeditions into uncharted lands.

Bolton thus challenged the notion that the westward movement across what was to become the United States illustrated a pioneering spirit that other nations lacked: "Everywhere contact with frontier environment and native peoples tended to modify the Europeans and their institutions. This was quite as true in the Latin as in the Saxon colonies" ("Epic" 453). This sense of multiple borderlands bolstered his view of America as a disputed hemisphere rather than a set of fixed national entities; he treated contested lands without imposing isolationist paradigms and recognized historical parallels that appear all the more radical for having been neglected for so long: "In Saxon America the story of the 'struggle for the continent' has usually been told as though it all happened north of the Gulf of Mexico. But this is just another provincialism of ours. The southern continent was the scene of international conflicts quite as colorful and fully as significant as those in the north" ("Epic" 453).

While Bolton's research on Florida and northern New Spain detailed Spanish contributions to what would become the U.S. national heritage, his greater ambition was to establish an academic legacy that would engage American history from the North Pole to the Tierra del Fuego. To this end, he concludes "The Epic of Greater America" with a list of suggested research topics meant to incite interest beyond the parochialisms of the nation-state:

> Who has written the history of the introduction of European plants and animals into the Western Hemisphere as a whole, or the spread of cattle and horse raising from Patagonia to Labrador? Who has written on a Western Hemisphere scale the history of shipbuilding and commerce,

mining, Christian missions, Indian policies, slavery and emancipation, constitutional development, arbitration, the effects of the Indian on European culture, the rise of the common man, art, architecture, literature, or science? Who has tried to state the significance of the frontier in terms of the Americas? (474)

In his call for hemispheric research, Bolton sketches a network of cultural, political, and biological filiations that Alfred W. Crosby would later term "the Columbian exchange." Indeed, in the almost eight decades since Bolton's address, many of these topics have been explored by scholars in a variety of disciplines. Bolton directed more than a hundred PhD dissertations at the University of California at Berkeley, thereby promoting a more ample Americanism through his own students.

In what follows, I consider several scholarly studies of hemispheric history and culture impelled by Bolton's inclusive vision of the Americas—a vision that underlies three interrelated fields of study: (1) Latin American history; (2) U.S. Hispanic border studies, and (3) Inter-American studies. All of this leads to my integration of a Boltonian perspective into two comparative courses I developed for a new "Literature of the Americas" curriculum at the University of Mary Washington in Fredericksburg, Virginia.

Latin American History

Bolton implemented his synchronic vision in "History of the Americas," a course he taught at UC Berkeley from 1919 until his retirement in 1944. Designated "History 8 A-B," the yearlong survey explored the colonial era in the first semester and the national periods in the second. Bolton designed the course to form the macroscopic component of a multilayered approach meant to coexist with offerings in national and regional histories, thus providing—not unlike European history courses—a fuller reflection of the interplay of local constituencies within a distinct world civilization (see Magnaghi 1998). It was so popular with students and faculty alike that in 1928 he published the course syllabus as *History of the Americas: A Syllabus with Maps*.

Although academic interest in the hemispheric model had waned by the mid-1960s, early on in his career Bolton had inspired many scholars (with a considerable number of his students among them) to follow in his footsteps. Like Bolton, these scholars saw beyond the clichés of Spanish cruelty and religious intolerance popularized in the United States by the publication of William H. Prescott's *History of the Conquest of Mexico*

(1843) and *History of the Conquest of Peru* (1847), and revived during the Spanish-American War. One of Bolton's most important students was Irving A. Leonard, whose *Baroque Times in Old Mexico: Seventeenth-Century Persons, Places, and Practices* (1959) is one of the most successful scholarly examinations of colonial New Spain by a U.S. historian, and is perhaps especially notable for disrupting the opposition between U.S. liberal democracy and Mexican autocracy. Other advocates of Bolton's Hispanic focus wrote histories of viceregal administration—for example, Arthur Scott Aiton's *Antonio de Mendoza: First Viceroy of New Spain* (1927) and Arthur Franklin Zimmerman's *Francisco de Toledo: Fifth Viceroy of Peru, 1569–1581* (1938)—while in *Academic Culture in the Spanish Colonies* (1940), John Tate Lanning recounted an illustrious history of Spanish education in the Americas, including the founding of ten major universities and the conferring of 150,000 degrees, many of them by institutions considerably older than Harvard (1636) and Yale (1701) (for example, the Universidad Nacional Autónoma de México in Mexico City and the Universidad Nacional Mayor de San Marcos in Lima, both founded in 1551). These narratives avoid both the lure of the "black legend" and the temptation to romanticize Amerindian empires as they trace the creation of new Hispanic American societies (see also Hanke 1947 and Keen 1985).

Border Studies

While many of Bolton's followers specialized in Latin America, some expanded upon his work on North American border history. In "The Significance of the Spanish Borderlands to the United States" (1976), Donald E. Worcester uncovers an Iberian legacy in the United States in areas ranging from agriculture and architecture to literature and the law. In addition, he shows that Hispanic immigration into the United States during the twentieth century follows colonial patterns of movement across Anglo-Hispanic borders. Américo Paredes, a pioneering figure in Chicano studies, has expanded the borderlands field as a poet, folklorist, and historian of greater Mexico. In his seminal *A Texas-Mexican Cancionero: Folksongs of the Lower Border* (1976), he transcribes and translates musical traditions from the lower Rio Grande border into English and details the creation of *corridos*, *danzas*, and other genres along with the social customs guiding their performance. The folksongs trace a history of the Texas-Mexican heritage from the colonial period (the 1750s) through the U.S.-Mexican War and up to the 1960s.[2]

Bolton's closest protégé in border studies, however, is David J. Weber. In *The Spanish Frontier in North America* (1992), Weber, like Worcester and Paredes, recognizes that

> [a]lthough the United States has always been a multiethnic society, most general histories of the nation have suggested that its colonial origins resided entirely in the thirteen English colonies. In American popular culture, the American past has been understood as the story of the expansion of English America rather than as the stories of the diverse cultures that comprise our national heritage. (5)

Thus, although Weber limits his North American inquiry to areas north of Mexico, his book helps to decenter the American national narrative from its presumed roots in the Anglophone northeast. In addition, Weber recognizes that "[a]s they sought to demonstrate Spain's lasting contributions to America, borderlands scholars had frequently failed to make clear that Spaniards achieved many of their 'successes' at agonizing cost to Indians" (359; see also *The Mexican Frontier* and *Myth and History*). Unlike Bolton, who seems unable to recognize racial differences within the Spanish borderlands, Weber does not regard Native Americans and *mestizos* as irrelevant historical entities or merely as victims of the U.S. genocidal march; rather, he demonstrates that they are participants in the networks of exchange and migration that characterize frontiers.[3]

Besides mapping the Spanish colonial presence in U.S. history, Weber also discusses the reinvention of this presence by Chicano and border historians. For instance, he describes how the myth of Aztlán is used to serve Hispanic intellectual and political ends in the 1960s. Identifying with its Indian rather than Spanish origins, the Chicano movement utilized the image of an Aztec homeland before their migration south to Tenochtitlán (now Mexico City) as a way of reclaiming an indigenous Mexican origin within the United States. In *Borderlands/La Frontera: The New Mestiza* (1987), Gloria Anzaldúa, another key contributor to border studies, uses the image of Aztlán to connect the pre-Hispanic goddess of birth and death, Coatlicue, to the Chicana feminist movement. Anzaldúa combines Nahua, Mexican, and U.S. Anglophone linguistic and cultural elements into a new border identity, a *mestizaje* that escapes the trappings of Western binary reason. Reinvented through the greater American frameworks of these scholars, Bolton's definition of the Spanish borderlands as an academic subject both prefigures and affirms U.S.-Latino studies.

Inter-American Studies

Unlike Bolton's legacies in Latin American history and border studies, his model for a hemispheric history has been largely ignored. A notable exception is the 1964 collection *Do the Americas Have a Common History? A Critique of the Bolton Theory*. Edited by Lewis Hanke, a supporter of Bolton, the book was the first to provide an in-depth analysis of his pan-Americanism. It contains Bolton's 1932 address, as well as numerous rebuttals, including a vitriolic response by the Mexican historian and philosopher Edmundo O'Gorman. In "Hegel y el moderno panamericanismo," which originally appeared in 1939, O'Gorman denounces Bolton's emphasis on material development as a marker of historical progress and his failure to recognize the important role of culture and religion in Latin America. The Bolton thesis, O'Gorman claims, is a "fallacious illusion" guided only by a "geographical hallucination" (107). Indeed, Bolton's crude developmental model, which reflects U.S. foreign policy as it was articulated in 1933 by Franklin Roosevelt, causes Bolton to misconstrue Latin America completely: "Truly, Professor Bolton with his well-intentioned, leveling vision, and with much of what—for causes baffling all reason—is called nowadays the 'Good Neighbor' policy, presents us with an inhuman history, an ample chronicle of an enormous organism indifferent to its salvation or to its perdition" (105). For O'Gorman, Bolton's history is nothing more than academic imperialism, an act of intellectual hegemony disguised as hemispheric cooperation. Because Bolton's sweeping approach fails to address Latin American cultural aspirations, O'Gorman argues that Bolton fails to establish a unified history based upon "that spiritual complex which gives body to a historical entity" (105).

Yet, if on the one hand O'Gorman identifies some of Bolton's methodological and cultural blind spots—particularly his ignorance of how indigenous issues shape *mestizo* identities in Latin America and his failure to examine inter-American diplomacy critically—on the other, O'Gorman's response appears to be generated as much by his own opposition to U.S. foreign policy as by Bolton's hemispheric model. In an attempt to counter an academic analogue of the "Good Neighbor" policy, O'Gorman embraces a Hispanic essentialism that presumably will prevent Latin American history from being subject to a Hegelian measure of progress that O'Gorman believes is an extension of U.S. imperialism.

Paradoxically, O'Gorman's *The Invention of America: An Inquiry into the Historical Nature of the New World and the Meaning of Its History*, delivered as a series of lectures at Indiana University in 1958 and published in a revised and expanded book form in 1961, legitimates the very comparability between the United States and Latin America that he dismisses in his earlier response to Bolton, and it does so by establishing the European imagining of the so-called New World as a point of origin for inter-American history:

> If one ceases to conceive of America as a ready-made thing that had always been there and that one day miraculously revealed its hidden, unknown, and unforeseeable being to an awe-struck world, then the event which is thus interpreted (the finding by Columbus of unknown oceanic lands) takes on an entirely different meaning, and so, of course, does the long series of events that followed. All those happenings which are now known as the exploration, the conquest, and the colonization of America; the establishment of colonial systems in all their diversity and complexity; the gradual formation of nationalities; the movement toward political independence and economic autonomy; in a word, the sum total of all American history, both Latin and Anglo-American, will assume a new and surprising significance. (45–46)

Invented rather than discovered, America is marked by a Eurocentric worldview that structures histories of colonization, revolution, and nation building. O'Gorman sees a complementary relation between New World cultures even as he accounts for a stark contrast between Anglo- and Latin America. Inquiring into the meaning of the history of the hemisphere, he maintains that Iberian America is the product of a transplanted European culture: The rules governing church and state, including the layout of cities and the system of social privilege, show Spanish adherence to "a sort of historical mimetism of Europe" (143). By contrast, Anglo-America is characterized by a dynamic process of transformation epitomized by frontier individualism and the U.S. Constitution. In essence, the United States' "desire for historical autonomy" gave birth to a pioneering spirit that resulted in the establishment of a radically new society separate from European civilization (143). O'Gorman concludes his study by interpreting the two Americas as complementary aspects of a hemispheric history:

> It was the Spanish part of the invention of America that liberated Western man from the fetters of a prison-like conception of his physical world, and it was the English part that liberated him from subordination to a Europe-centered conception of his historical world. In these two great liberations lies the hidden and true significance of American history. (145)[4]

Other scholars refined Bolton's comparative methodology within their own academic fields. Arthur P. Whitaker, a historian of Latin America involved in the History of America Program sponsored by the Pan-American Institute of Geography and History, supported Bolton's common history thesis, but only for the colonial period (see "The Americas"). Inspired by Bolton's hemispheric connections, Mexican philosopher Leopoldo Zea sought "to achieve a full cultural interpenetration between the two Americas" ("Interpenetration" 539) in the hope that, despite the imbalances of U.S.-Latin American relations, a transnational synthesis could place "the idea of America, i.e., that which is common to our two Americas" (543) within a single philosophical framework, for, while "diversity establishes these Americas as contradictory... the contradiction is not so great as to make impossible the inclusion of the forces in one system" (538; see, also, *The Role of the Americas*).[5]

More recently, interest in the hemispheric model has surfaced in American and Latin American studies as part of a shift to more global disciplinary contexts.[6] Felipe Fernández-Armesto, J. H. Elliott, and Jorge Cañizares-Esguerra, for instance, have adapted Boltonian Americanism to their comparative New World histories. In *The Americas: A Hemispheric History* (2003), Fernández-Armesto first treats the vast and complex indigenous and Spanish American civilizations of the Americas and then adds the United States as another "civilization" with its own distinctive features, thus denaturalizing the primacy of the United States by making pre-Anglophone empires normative. Elliott's *Empires of the Atlantic World: Britain and Spain in America, 1492–1830* (2006) revises Bolton's hemispheric agenda by limiting Bolton's comparative project to Spanish and British colonial American traditions, for "[g]iven the number of colonizing powers... and the multiplicity of the societies they established in the Americas, a sustained comparison embracing the entire New World is likely to defy the efforts of any individual historian" (xvi). And in *Puritan Conquistadors: Iberianizing the Atlantic, 1550–1700* (2006) Cañizares-Esguerra provides an account of evangelical traditions and their shared role in the colonization of North and South America, showing how, despite the religious differences between Catholics and Protestants (one of the most commonly rehearsed differences between Anglo- and Hispanic America), the exorcising of "demonic forces" from the New World created a common foundation for European conquest.

Toward a Hemispheric Literary Paradigm

Scholars are also increasingly endorsing a hemispheric framework in the realm of American literary studies. Most notable is the critical anthology *Hemispheric American Studies* (2008), edited by Caroline F. Levander and

Robert S. Levine, which aims to "chart the interdependencies between nations and communities throughout the Americas" (6) and to raise awareness of "the importance of doing literary and cultural history from the perspective of a polycentric American hemisphere with no dominant center" (7). The essays explore the promises and perils of hemispheric approaches to, among others, early American, African American, Asian American, and Southern U.S. literatures and cultures. The volume includes fifteen varied essays by prominent U.S. academics, but its outlook remains limited by the linguistic, geographic, and disciplinary boundaries and biases of English and American studies departments, for whom "hemispheric" often means little more than an expanded national terrain.

Because the transnational inquiry is taking place mostly in national literature departments, scholars need to question how the academy conceives of and supports a branch of research that began with books written primarily by scholars in comparative literature.[7] In 1980, Earl Fitz, a pioneering inter-Americanist who had just created six courses in the field at Penn State University, made the following forecast: "It is our contention that inter-American literary studies, naturally of a comparative nature, will prove themselves to be a major trend of the near future, one which will eventually establish itself as a permanent and vital part of every comparative literature department and program in the country" (10). Fitz, of course, was only half right. While academic inter-Americanism is on the rise in American literary studies, the majority of academic positions and fellowships designated as "Literature of the Americas" and "Transnational American Studies" are housed in English departments, where Anglophone literature holds precedence over other hemispheric literatures. As a result, most programs remain monolingual and national in scope, and adopt a historically anachronous approach that places British before Spanish American culture. At best, this method uncovers a hemispheric imaginary at work in Anglophone (including Latino) U.S. fiction, while failing to decenter the nation. Considering the extent to which the new hemispheric literary studies is an imperialistic strategy, Claire Fox has recently characterized the general disregard for "Literature of the Americas" as "another name for half-hearted literature in translation courses; for neoliberal 'dialogues' among English, foreign languages, and ethnic studies; and for the fashionable Anglo-American *mea culpa* that strives to think globally while controlling the terms, and language, of the debate" ("Comparative Literary Studies" 871).

We need to inquire into the motivations of Americanists to go hemispheric at this particular time, but without retreating to an isolationist deadlock that undermines the vigorous study of the Americas. While growing increasingly sensitive to the politics of our own research, we might avoid a two-way exceptionalism that obfuscates the wider latitudes of transatlantic history and intercultural exchange across the hemisphere. For, as is illustrated by the Bolton-O'Gorman debate, one of the things that unites the Americas seems to be the tendency toward formulations that ignore and/or repress inter-American bonds. Thus, while I remain critical of the Bolton thesis for its underlying political misconceptions about the cooperative role of the Americas, I find it impossible to ignore the emergence of a valid site of interdisciplinary research. I invoke Bolton as a pioneer of inter-American studies in order to affirm that what strikes me as a bold trajectory might guide further comparative inquiry.

To the extent that it is both comparative and historically grounded, Bolton's methodology seems to me an important precondition for the creation of a coherent trans-American studies project. Following his model in a modified form that includes European, African, and indigenous histories across the Americas, we might envision a comparative American literature that is geographically and culturally correct (that is, one that includes Canada, South America, and the Caribbean, as well as the United States). Accordingly, this multinational and multilingual field should allow for multiple points of entry between and among New World traditions. It should support comparative analysis of both the United States and its ethnic and border cultures, and of other hemispheric literatures within trans-Atlantic and trans-Pacific contexts (a type of research admittedly more characteristic of Latin Americanist scholarship, which quite often deals with Hispanic and Lusophone traditions across nations and regions). Under this literary paradigm, hemispheric study might include comparisons across nineteenth-century Brazil and Argentina, or explore the revolutionary contexts of Québécois and Cuban fiction from the 1960s.

Without the comparative American perspective, it is doubtful that the U.S. academy can ever establish "Literature of the Americas" programs that actually address—and, ideally mitigate—the ideological divide between the United States and its hemispheric neighbors, or, for that matter, between Latin American nations that are invested in their own isolated histories. No doubt, this huge body of work calls for collaboration among scholars working in different literary traditions and periods of specialization. Yet, this discord should not keep us from pursuing a course of study that we know to be right. Just as inter-American study is best seen as a collaborative project, so should an Americanist, by definition at least, be a comparatist.

I conclude this sketch of Bolton's legacy by turning briefly to my own curricular development as someone trained in comparative literature. At the University of Mary Washington, I have created two new courses inspired by Bolton's historiography: "New World Writing in the Colonial Period" and "Literature and Nation Building in the Americas." The first course surveys writings from the colonial period in a hemispheric context. Guided by a series of inter-American juxtapositions, it traces authorial strategies used to negotiate colonial New World identities. Selections range from accounts by European explorers, such as Christopher Columbus and Jacques Cartier, to narratives of independence written by Thomas Jefferson and Simon Bolívar. Topics covered include precontact cultures, the literature of the encounter, the psychology of the conquest, race and transculturation, Amerindian and African slavery, and the influence of the Enlightenment on American revolutions. The second course examines literature in relation to the establishment of independent republics starting in the late eighteenth century. Topics covered include the rise of national and pan-American cultural ideologies (reflected through the intellectual manifestos of Ralph Waldo Emerson, Thomas D'Arcy McGee, and José Martí), Native American dispossession, race and miscegenation, democracy and dictatorship, New World plantation cultures, the experience of industrial development, and the border fiction written in the wake of the U.S.-Mexican War by María Amparo Ruiz de Burton and others.

Through a series of juxtapositions that places the literatures of North America, South America, and the Caribbean side by side, these courses present students with a wide variety of texts and ask them to examine established materials in new ways; the shock of the unfamiliar America competes with a more familiar America in ways that position Amerindian, Francophone, Hispanic, and Lusophone texts as intrinsic to a New World tradition. The courses highlight inter-American themes through three types of interconnection: (1) parallelism—writings influenced by common histories, such as the "discovery" epistles of Hernán Cortés and Amerigo Vespucci; (2) intertextuality—authors that influence each other across hemispheric borders, such as Walt Whitman's impact on José Martí; and (3) inter-Americanism—authors whose themes and cultural positioning traverse the Americas, such as Joel Barlow's incorporation of the Inca in *The Vision of Columbus* (1787) and Henry Wadsworth Longfellow's depiction of Acadian deportation in *Evangeline: A Tale of Acadie* (1847). While selections are based upon parallel and overlapping histories of contact, conquest, colonization, revolution, and nation building, my approach emphasizes the cultural and political specificities that make New World identities plural and often contradictory.

Now that comparative literature has expanded to the Americas, it is up to scholars to embrace the hemispheric paradigm in practice and not just in theory. Most of all, it is imperative that the discipline retain the agonistic basis of its existence. In the words of Franco Moretti, comparative literature must continue to be "a thorn in the side, a permanent intellectual challenge to national literatures... If comparative literature is not this, it's nothing" (68).

Notes

1. For a different reading of Bolton's title, particularly his misuse of the term "epic" to denote a non-national narrative, see Truett.
2. Paredes's work ranges from musicology and ethnography to themes of machismo and anti-Mexican sentiment in the United States. See, especially, *Folklore and Culture on the Texas-Mexican Border* (1993). Saldívar argues that Paredes is also a forerunner of interamerican studies: "Paredes ordinarily sees the national culture or political event as a local inflection of a transnational phenomenon that can only be read according to a hemispheric dialectic of similarity and difference" (9–10).
3. For an interpretation of Latin American frontiers, see Zavala. A good starting point for tracing Bolton's impact on borderlands historians is *Bolton and the Spanish Borderlands*, ed. Bannon.
4. Dussel (27–36) denounces O'Gorman's conception of a New World tabula rasa in order to enact a universal history in the Americas. In Dussel's view, O'Gorman's Eurocentric perspective fails to account for the brutal underside of a modernity that began in 1492, namely the genocide and dispossession of Amerindian communities.
5. The History of America Program, sponsored by the Pan-American Institute of Geography and History, was partly inspired by "The Epic of Greater America." For a brief summation, see Hanke, *Do the Americas Have a Common History?* 30–38. The institute published a Spanish translation of Bolton's address, titled "La epopeya de la máxima América," See also Zea's *The Role of the Americas in History* (1992).
6. Founded in 2003, the International American Studies Association (IASA) has moved most emphatically toward a global framework. See founding IASA president and guest editor Djelal Kadir's Introduction to the *PMLA* special issue *America: The Idea, the Literature*, which he also edited. For scholarship that traces the shift to a hemispheric model, see *Critical Perspectives and Emerging Models of Inter-American Studies*, a special issue of *Comparative American Studies*, ed. Claire F. Fox, and *Hemispheric American Literary History*, a special issue of *American Literary History*, ed. Caroline F. Levander and Robert S. Levine. More recently, the 2007 National Endowment for the Humanities seminar titled "Toward a Hemispheric American Literature" (directed by Rachel Adams and Caroline F. Levander).
7. See, for example, Kutzinski, Ramon Saldívar, both studies by Zamora, and Fitz *Rediscovering the New World*. See also the collections edited by Chevigny and Laguardia and Pérez Firmat. For a comprehensive list of primary and secondary sources, see Fitz, *Inter-American Literature and Criticism*.

CHAPTER 2

Bad Neighbor/Good Neighbor: Across the Disciplines toward Hemispheric Studies

Caroline F. Levander

Significant challenges confront U.S. Americanists who want to adopt transnational, inter-American, or hemispheric frameworks in their scholarship. Hemispheric models, as Latin Americanists in the United States and Latin America have observed, can run the risk of reproducing the totalizing, neoimperialist structures that American studies practitioners may be seeking to work against—such models inadvertently can take part in the economic and cultural domination of Latin America by the United States in the name of globalization. As Sophia McClennen provocatively asks: "Inter-American Studies or Imperial American Studies?" McClennen's apprehension that hemispheric or inter-American studies is "the latest variation on the Monroe Doctrine of patronizing Latin America" (394) reflects the concern of Latin Americanists, as well as those teaching in language departments, that hemispheric models will reorganize institutional resources and curricular structures in ways that will further marginalize already struggling language and area studies disciplines.

But the challenges hardly end there. American history, literature, and cultural studies scholars engaged in the hemispheric turn that has been underway since the 1990s can, likewise, run the risk of mystifying important differences between the historical experiences of Canada, Latin America, and the United States—of failing to recognize that, for example, the idea of the nation-state has different connotations in Latin America, where national boundaries are often seen as a protection against U.S. cultural, economic, and military expansion, than in the United States, where these same national borders are often approached by scholars as artificially constructed, "imagined," or contested. These scholars can, on occasion, tend to "discover" what some Latin Americanists may have already documented, possibly overlooking some of the vast scholarship already undertaken that considers related questions but that is published in non-U.S. venues or in languages other than English.

The recent focus on the United States in a hemispheric context has generated some concern among U.S. scholars of minority cultures and area studies scholars as well. Some U.S. African American studies scholars, for example, have voiced apprehension that hemispheric American studies tends to be dominated by Latino/Latina studies and Latin American studies, and that this focus on a brown/white story effectively makes the place of African American studies unclear in hemispheric studies. Scholars of gender studies have likewise noted that a hemispheric model directs primary attention to racial migration and slavery as a hemispheric system of oppression, thereby obscuring questions of gender oppression and patriarchy. Finally, hemispheric studies runs the risk, as Paul Giles and Alberto Morieras have observed, of simply replacing nationalist essentialism with hemispheric essentialism—of re-creating on a larger geographic scale master narratives that uphold the idea of hemispheric exceptionalism as opposed to national cohesiveness and, alternately, of obscuring the even larger-scale events of which discrete localities form a key part. In sum, this work can tend both to adopt too large of an interpretive framework—obscuring important distinctions within and between national traditions—and to adopt too small of an interpretive framework—obscuring important large-scale, global phenomena that link discrete parts of the globe into interlocking world systems.

Given such pitfalls, we might ask why some scholars have persisted in questioning rather than assuming the hegemony of the U.S. nation— why they have continued to shift the field imaginary away from a self-evident, U.S. nation-based frame of reference toward a geopolitical field of inquiry that highlights the points of overlap and convergence between nations and peoples. My task in the following pages is not to champion a wholesale intellectual embrace of hemispheric models for the study of American history, literature, religion, and culture; nor is my task to weigh in on whether or not certain disciplines (such as comparative literature or Spanish or Portuguese) have an intellectually privileged position from which to issue hemispheric edicts. Rather, I would like to suggest that there are some questions and lines of inquiry that are particularly well served by the adoption of a hemispheric framework. Scholars in humanities departments at many U.S. institutions have been asking these and other questions with increasing regularity and urgency in the last few years—for example: What is the utility and durability of the idea of "nation" in a global era? Does it make sense to retain "nation" as a defining feature of disciplinary rubrics? How do we categorize, analyze, and conceptualize our various field imaginaries once we revisit the geographic assumptions that have so long shaped disciplinary parameters? Rather than joining the debate about whether those who teach and study the United States within hemispheric contexts are unilaterally engaging intellectually in "good" or "bad" neighborly behavior—in other words, whether such work currently taking place across the cognate disciplines of the humanities and humanistically inclined social sciences refutes or extends imperialist logics, elucidates or strategically overlooks some racist and sexist discourses, etc.—I would like to shift the terms of the conversation to consider the knowledge that such scholarship has recently generated, to comment on the different kinds of questions that this work has enabled, and to highlight some of the enduring questions to which this scholarship has contributed. Doing so interjects into the current conversation a long view that highlights what we gain and what we give up when scholars undertake the task of teaching and studying the Americas in ways that do not presume but rather explore the shape and texture of our object of inquiry.

First, it is worth remembering that the strand of scholarly inquiry that has produced much, often contentious, debate in the last few years isn't particularly new, despite what many of its champions and detractors tend to think. Rather, consideration of the hemisphere as an interlocking, internally fractured conceptual frame goes back well over seventy years, to UC Berkeley historian Herbert E. Bolton's seminal 1932 presidential address to the American Historical Association (AHA), "The Epic of Greater America." In what became known as the "Bolton thesis," the AHA president, as Antonio Barrenechea observes, challenged historians to recognize that each individual national story has "clearer meaning when studied in the light of the others; and that much of what has been written of each national history is but a thread out of a larger strand."[1] In words uncannily similar to those used by Bolton, Janice Radway in her 1998 presidential address to the American Studies Association, critiqued the ASA's use of "American" to designate an organization whose more accurate field of inquiry should make it "The United States Studies Association" (448).[2] Bolton's controversial thesis subsequently forced scholars to think about the commonalities and divergences in the colonization, conquest, and development of the Americas, generating Lewis Hanke's anthology *Do the Americas Have a Common History?* (1964), among other responses. In the 1950s and 1960s, the work of Antonello Gerbi and Hans Galinsky propelled inter-Americas scholarship in Europe, while scholars like Bell Gale Chevigny and Gari Laguardia, Vera Kutzinski, and Lois Parkinson Zamora moved inter-Americas scholarship forward in the U.S. academy of the 1980s. Thus, when, in 1990, Gustavo Pérez Firmat asked, "Do the Americas have a common literature?," he was contributing to a sustained, albeit episodic, critical conversation that had been ongoing, intermittently, for over half a century. Firmat's question, in turn, generated healthy debate in literary criticism in the early 1990s. Such a question has yielded important discoveries about a common critical heritage that joins the diverse literatures of North, Central, and South America.[3]

Bolton was not only asking scholars working on the history of the United States and the Americas to see disparate parts of the hemisphere as parts of an interlocking whole—he was, on a more fundamental level, asking them to reckon fully with the atypical history of nationalism and nation formation in the Americas. As Don Doyle and Marco Antonio Pamplona, among others, have illustrated, it is important to remember that nationalism functioned differently in the Americas than in other parts of the globe, in part, because the very ideas of nation and nationalism in the Americas have historically not had as firm a hold as they have in Europe, the Middle East, Africa, or Asia. As a result, the lively discussion of nationalism that has been ongoing over the last twenty-five years has largely ignored the Western Hemisphere and has generated models of nationalism that do not accommodate the Americas. This oversight is exacerbated by the fact that, as Walter Mignolo observes, the logic of coloniality operative in the Western Hemisphere is one in which racial identifications bleed over and often trump national demarcations. Therefore, not only are racial affiliations across geography dominant in the Americas, but also pluralism and the newness of American nations meant that ethnic nationalism was, of necessity, a less important adhesive source of nationhood in the Americas than in Europe. Anticolonial and less fueled by ethnonationalism, independence movements in the Americas were also less frequently challenged by separatist movements once nations were formed (a key exception being the U.S. Civil War). Yet American nations experienced tremendous ethnic tensions within their borders. These intranational ethnic tensions, coupled with the fact that ethnonationalist conflicts between American nations were relatively rare, compared to the rest of the globe, meant that the nations comprising the Americas have consistently defied nationalist paradigms and have therefore tended to be overlooked in discussions of nationalism. Therefore, we must exercise care when we apply critical rubrics of nationalism to the nations of the Americas, because we can run the risk of obscuring key elements of nationalism and national identity among American nations.

Not only have nations throughout the Americas not tended to conform to nationalist paradigms, but the idea of nation has also been challenged, even—or rather, especially—in the case of those countries, like the United States, where the idea of nation can seem, in retrospect, to be historically transcendent, ever-present, and largely ubiquitous. Therefore, it is not only the shape and texture of Americas' nationalisms to which we must attend when we talk about the genesis, history, and interconnections of nations in the Americas, but also the advent and genesis of the idea of nation itself. As Walter Mignolo and Edmundo O'Gorman, among others, have observed, the ideas of America and Latin America were invented rather than found. Even in the case of that most notorious of American nations—the United States—it is important to remember that it is only from our present vantage point that the U.S. national hegemony of which many Latin Americanists are suspicious seems inevitable and omnipresent. Founding fathers, for example, were acutely aware that, as John Adams wrote, "the character of Gentlemen in the four New England Colonies differs as much from those in the others as that of the Common People differs, that is as much as several distinct Nations."[4] Connecticut delegate William Williams in August 1776 observed that "the ideas of North and South (or as now properly called East and West) [were] as wide as yer Poles, [with] such clashing & jarring Interests [and] such diversity of manners" that he "little expect[ed] any permanent Union."[5] Even after the Articles of Confederation were ratified in 1781, many congressional leaders doubted that national governance would last because states' rights overpowered any fledgling national affiliation that occupants might feel. Virginia delegate James Madison, for example, recommended to his state legislature that it should "presume that the present Union will but little survive the present war."[6] James M. Varnum, delegate from Rhode Island, predicted in 1781 that "the time is not far distant when the present American Congress will be dissolved, or laid aside as Useless, unless a Change of Measures shall render their Authority more respectable."[7] In other words, in the immediate wake of its inception, the very viability and longevity of the nation that has subsequently been understood to wield disproportionate power throughout the hemisphere was not assumed or expected by the nation's founders, let alone by those who resided within its boundaries.

Over the next hundred years, questions about the nation's longevity and durability were not entirely eradicated, despite exertions of U.S. muscle, and the idea that the United States was or could ever be fully invented as a nation continued to be refuted by leading political figures. According to the famous diarist and New York lawyer George Templeton Strong, for example, the "bird of our country is a debilitated chicken, disguised in eagle feathers. We have never been a nation; we are only an aggregate of communities, ready to fall apart at the first serious shock and without a center of vigorous national life to keep us together." His observation in 1854 was only confirmed with secession, and Strong concluded that Americans were "a weak, divided, disgraced people, unable to maintain our national existence...impotent even to assert our national life."[8] The Civil War, according to a dominant story about US national unification, is what made these United States crystallize into a single national entity that has the look of our modern national government. At this particular moment of national consolidation, Charles Sumner brings this modern concept of nation to bear retroactively on the messy past events out of which the modern concept of nation became operative and, expressing the views of many in his essay "Are We a Nation?," concluded that "even if among us in the earlier day there was no occasion for the word Nation, there is now. A Nation is born."[9] However, Thomas Wentworth Higginson continued to draw an important distinction. In "Americanism in Literature," he observed that Americans tended "to say that the war and its results have made us a nation, subordinated local distinctions, cleared us of our chief shame, and given us the pride of a common career." But the war had clearly not, in Higginson's estimation, created a uniform national identity—a pervasive feeling within "all persons among us" of "Americanism."[10] Soldiers purportedly fighting for the "great cause" of national unity were often keenly aware of the hypocrisy of nationalizing rhetoric. As one Rhode Island private wrote disgustedly, "officers triing [sic] to make young men believe that they are fighting for the Union [are] false as hell." "Northern fanaticks [sic] do not care a dam [sic] for the union or the country," he concluded.[11] Three decades later, writers were still cognizant of how the ethnic and racial diversity within the country counteracted the consolidation of a unified national ethos. As one writer observed in 1898, "we are still in the process of forming, as a nation."[12] As these comments suggest, on the ground, many Americans contested and doubted the durability of the idea or invention of the United States as a nation. Therefore, when we project current ideas of the U.S. nation backward in time, we unwittingly present the creation of nation as a clear, unbroken, transhistorical phenomenon rather than as a concept that was contingent, contested, and uncertain well into the late nineteenth century. Furthermore, we fail to remember that dissent over a national ethos can be, and often is, a fundamental part of national articulation and affirmation as much as a challenge to national cohesion, particularly in a nation that embraces Lockean models of consensual governance.

Once U.S.-focused scholars approach their object of study with the partial, fractured, and indeterminate nature of the early nation clearly in mind, a dynamic, internally riven community comes into view—a dissonant as well as consonant collection of peoples and views. Therefore, it comes as no surprise that portions of this United States might at times have had more in common with other parts of the hemisphere than with the disparate regions within its borders. The idea of a well-formed, uniform U.S. nation springing full-fledged from Puritan colonies is invariably anachronistic, as we have seen. Imposed retrospectively on what was a much messier set of debates on nation, union, and geopolitical futurity than we have generally allowed, the idea of a fully formed early U.S. nation may work to highlight the fissures, conflicts, and power disparities *between* nations, but it does not register the uncertainty, provisionality, and tenuousness of the idea of nation within U.S. borders. Nor does it acknowledge how the internal variations and degrees of affiliation that diverse constituents felt and experienced along a continuum of citizenship in the United States at times created more common ground among subsets of U.S. citizens and inhabitants of other nations than among other occupants of the United States.

The organization of many humanities fields along national lines has tended to mitigate against studying and teaching the Americas in ways that emphasize communal affiliations that cross or ignore national lines. English and history departments, for example, have historically tended to be organized along "American" and "British" tracks, thereby tacitly reinforcing the hegemony of national differences. Even comparative literature departments too often tend to compare the literary traditions of different nations, thereby tacitly reinforcing the centrality of national heritage, while language departments are often conceptualized around geographic loci of shared linguistic cultures. But important strands of scholarship have emerged at the interstices of these well-established disciplines and have worked against the grain of well-worn institutional rubrics, unearthing the histories, cultures, and literatures of those constituencies that develop in the overlaps and points of convergence between and across nations.

Take, for example, the experiences of newly freed blacks in Louisiana. Taking as her subject postslavery Louisiana and Cuba, historian Rebecca Scott illustrates how these two communities were jointly involved in thinking through the legal, economic, cultural, and social implications of freedom. As Scott illustrates, the urgent question confronting these two communities was not one of freedom versus enslavement, as it had been for U.S. Northerners contemplating slavery within the United States. Rather, the question being tested was one of tolerable degrees of freedom for blacks. While Louisiana's blacks continued to lose rights at the end of the century, across the Gulf of Mexico, voter rolls swelled. Such analysis of freedom across the national boundaries of slave-holding communities reveals the limited degrees of freedom that new U.S. citizens actually enjoyed relative to free blacks throughout the Americas. Scott's insights into how former slaves in two of the Western Hemisphere's most important slave societies tried to breathe substantive life into the idea and experience of freedom depends on analysis of the political, social, and economic worlds of Cuba and Louisiana after slavery. But it also depends on seeing Louisiana's newly freed slaves as more integrally connected to their Cuban counterparts than to fellow black citizens in Boston or New York.

Once we see that the U.S. slave system operated not only within the nation's boundaries or by way of the "black Atlantic" routes that Paul Gilroy has so powerfully described, we can begin to see U.S. slavery as circulating within a Gulf system that places U.S. slaves and slaveholders within a dynamic economic and cultural terrain—what Matthew Guterl terms an American Mediterranean of sorts. As scholars such as Guterl, among others, have recently illustrated, U.S. slavery was a hemispheric, as well as national and transatlantic, phenomenon. Such work offers not so much a comparison of different national slave systems, but rather an analysis of slavery as a hemispheric organization and arrangement. *Look Away!: The U.S. South in New World Studies* begins its consideration of the U.S. South in relation to Latin America and the Caribbean with the observation that some of the major characteristics that mark the South as exceptional within the United States—including the legacies of a plantation economy and slave trade—are common to most of the Americas. Both center and margin, victor and defeated, and empire and colony, this New World South, as Jon Smith and Deborah Cohn illustrate, has much to teach us about postcoloniality, territoriality, and geopolitical formations. In *American Mediterranean: Southern Slaveholders in the Age of Emancipation*, Guterl gives us a prime example of what can be learned once we see the U.S. South not only as the southernmost tip of the United States, but also as the northernmost edge of a slave-holding American empire. After tracing the links that bound U.S. slave owners to the wider fraternity of slaveholders in Cuba, Brazil, and elsewhere, Guterl explores how the Southern elite connected—by travel, print culture, and even the prospect of future conquest—with the communities of New World slaveholders. In his analysis of how and why they invested in a vision of the circum-Caribbean, Guterl shows how this pan-American master class developed a shared consciousness that effectively relocated them within a circum-Atlantic topography that included the United States, the Caribbean, and Latin America.

As these few examples of the wealth of recent historical and cultural work on slavery in the Americas suggest, approaching the subject of U.S. slavery with a keen awareness of the atypical development of the idea of nation and nationalism in the Americas, as Bolton suggests, brings to light meaningful links that make the U.S. slave system part of a Mediterranean whole—the northernmost spoke of a multipronged circum-Caribbean wheel that includes Cuba and Brazil, among other regions. As the example of slavery so powerfully suggests, parts of one nation—in this case the United States—developed economic, social, political, and cultural bonds with regions throughout the Americas precisely because of shared interests that at times trumped or made of secondary concern citizens' unilateral commitment to the idea of national unity. Once we recognize this fact, we can begin to see how many of those events or authors that U.S. American studies scholars take as their subject can be located within networks that exceed or largely ignore national boundaries.

A hemispheric approach, then, may be worth adopting—not wholesale in order to uniformly assert or assume affinities among the histories and cultures of the New World, but rather strategically—when a nation's physical borders or the concept of the nation-state provide only partial rubrics for understanding the problem at hand. When we bear clearly in mind the fact that nations throughout the Americas are not only involved in a constant process of being made and unmade, just like they are in other parts of the world, but also share an unusual and largely atypical history when it comes to the phenomena of nationalism and the idea of nation, we can begin to see how this history may have helped give rise to an outpouring of cultural artifacts, narratives, works of art, and literatures that engage in a myriad of ways with the region's complex genesis and development.

Take, for example, those writers who we have historically identified as key figures in the development and rise of U.S. literary nationalism—writers like Walt Whitman or Nathaniel Hawthorne. Rather than writing of a nation fully formed, with a long-standing heritage that extends back to its colonial origins, as we often assume, such writers also contemplated the incipient nation's engagement with regions, communities, and cultures throughout the hemisphere and found in these roots, as much as in the nation's colonial past, the rudiments of a cohesive American literature. American literary scholars have tended to understand Whitman as forging a distinctive U.S. national heritage with his poetic celebration of the individuals who collectively epitomize a democratic American voice, but when approached with the nation's links to regions across the hemisphere clearly in mind, he can be seen to contemplate—and foreground—the importance of these interdependencies to the founding of a national literary tradition.

For example, in his 1883 letter to the city of Santa Fe, on the 333rd anniversary of its founding, Walt Whitman acknowledged that "we Americans have yet to really learn our own antecedents," which he asserts are "ampler than has been supposed." Whatever we might think of his anthropology, Whitman's continental vision belies the assumption—too often made by scholars—that strict racial lines delineate the topography of the nations comprising the Americas. It may be less the case that thinkers such as Bolivar and Jefferson necessarily thought about two Americas—"an Iberian America that extends to what are today California and Colorado [and] an Anglo-Saxon America [that] did not go further west than Pennsylvania," as Mignolo suggests, and rather that some thinkers, such as Whitman, grappled with what the complex overlay of peoples that have lived within national borders might mean for the idea of nation and national futurity. It is "a very great mistake," Whitman contends, to blindly accept the idea "that our United States have been fashioned from the British Island only and essentially form a second England only." Indeed "many traits of our future National Personality and some of the best ones, will certainly prove to have originated from other than British stock," Whitman concludes, and "Spanish character will supply some of the most needed parts" of the "American identity of the future." He explicitly turns to what was then the edge of the nation, rather than its centerpoint, to find the glue that will bind the disparate parts of the nation into a cohesive whole, and it is "the Spanish stock of our Southwest" that Whitman contends will provide a key element to the process of developing a national literature.[13] A decade later, Whitman was still engaged in the task of thinking through the composition of a national U.S. literary heritage, and in "Have We a National Literature?" (1891), he contends that, in its current state, American literature is at best in inchoate form. Quoting Margaret Fuller, Whitman agrees that the fact that "the United States print and read more books, magazines, and newspapers than all the rest of the world" does not mean that it therefore has a literature. To produce a "real representative National Literature" Whitman contends that "New England...and the three or four great Atlantic-coast cities, highly as they to-day suppose they dominate the whole, will have to haul in their horns." Because America is composed of "a huge world of peoples...and geographies—44 Nations curiously and irresistibly blent and aggregated in ONE NATION"—American literature must represent "this vast and varied Commonwealth" rather than confine its purview to "constipated, narrow and non-philosophic" "puritanical standards."[14] Whitman's reliance on the voices of those whose roots extended beyond the U.S. borders in his efforts to think towards a national literature, as Doris Sommer, Fernando Alegría, and Deborah Cohn, among others, have observed, resonated and created reactions throughout the Americas, and Spanish American authors like Pablo Neruda and Jorge Luis Borges both wrote against and appropriated Whitman's absorptive poetics in their efforts to generate Americas literary traditions.

Like Whitman, Nathaniel Hawthorne also found and generated creative inspiration and rich literary material at the edges of the nation's borders, as well as within the colonial heritage, that he often explicitly took as the subject of his fiction. Long recognized as writing out of the Puritan legacies of the U.S. colonial era, Hawthorne nonetheless found his initial creative inspiration in childhood stories about the nation's place and engagement in the hemisphere. His genesis as a writer and original creative inspiration, according to his son Julian, derives from his reading about the early nation's encounters with the West Indies. Recalling Hawthorne's lifelong fascination with the logbook that his father, Captain Nathaniel Hathorne, kept "during one of his voyages to and from the West and East Indies," Nathaniel Hawthorne's son emphasizes the pivotal role that these stories had on his father's literary vision. "This book," Julian recalls, "was in possession of the son" and "became the companion" of his "childhood and boyhood," providing the seeds for Nathaniel Hawthorne's literary imagination. As a child, Nathaniel was "in the habit of poring over it and making up many imaginative stories for himself about the events of the voyage," particularly the fight his seafaring father documents between his ship, *The Herald*, and pirates seeking to overtake the British East India Co.'s ship, *The Cornwallis*.[15] Scholars have tended to focus on the significance of Puritan history in the creation of Hawthorne's New England consciousness and such published works as *The Scarlet Letter*, *The Blithedale Romance*, and *The House of Seven Gables*. Yet his enduring fascination with stories of adventure that link New England to regions within Central and South America continue to be evident in writing that has historically occupied a marginal place in the Hawthorne canon—writing such as his *Life of Franklin Pierce* (1852). *Life of Franklin Pierce* was written with the explicit purpose of helping in the U.S. presidential bid of Hawthorne's lifelong college friend, Franklin Pierce. In this most overtly national and political of all of his texts, Hawthorne turns to the nation's shifting southernmost edge and analyzes in detail Pierce's involvement in the U.S./Mexico territorial conflict. In *Life*, Hawthorne's prose intermingles with Pierce's "Journal of his March from Vera Cruz," suggesting the extent to which, for Hawthorne, the political future of the United States and its writing were integrally bound to the nation's shifting outer and southern edges—the blurry boundaries where national affiliations did not hold and where national projects clashed. Like Whitman, Hawthorne emerges as a founding U.S. literary father whose conceptualization of national futurity involves looking forward and south as well as backward and east. And like Whitman, Hawthorne not only drew literary source material from throughout the Americas, but, in so doing, also generated important source material for American writers like Octavio Paz, whose 1956 play *La hija de Rappaccini* recapitulates central precepts of Mexican national identity through recourse to Hawthorne's "Rappaccini's Daughter."

As the examples of Hawthorne and Whitman suggest, many originators of U.S. literary nationalism as we now know it found a rich source of literary material in the nation's immersion in hemispheric contexts and consistent engagement with regions throughout the Americas. Just as important, authors to the south of the U.S. national border generated a rich literary tradition out of the complex ties that bound disparate regions on either side of this, at times seemingly arbitrary, national divide—writing that is as integral to conceptualizing American literary studies as is that of Whitman or Hawthorne. Nineteenth-century Cuban and Mexican writers such as Lorenzo de Zavala, Loreta Janeta Velazquez, and Olga Beatriz Torres, for example, composed important narratives that theorized the convergences and divergences among communities on either side of the U.S. southernmost national border, helping to shape the very hybrid American literature of which Whitman was in search. Their writing emerged out of their own multidirectional movements across national borders and their active leadership in determining these borders through their participation in such key political conflicts as the separation of Texas from Mexico (in the case of Zavala), the U.S. Civil War (in the case of Velazquez), and the Mexican Revolution (in the case of Torres). Lorenzo de Zavala (1789–1836) was a political leader in the newly established Republic of Mexico and the Mexican minister to France before traveling in 1830 across the United States with the goal of educating himself in democratic principles so that he might aid in the creation of a liberal national identity for the Mexican people. Written before Alexis de Tocqueville's *Democracy in America* (1835), which has been widely accepted as the first in-depth account of American democracy, Lorenzo de Zavala's *Viaje a los Estados Unidos del Norte América (Journey to the United States)* (1831) offers an in-depth analysis of "the manners, customs, habits, and government of the United States."[16] When Mexican President Antonio López de Santa Anna assumed dictatorial power in 1833, Zavala resigned his Mexican position and brought his political leadership skills to the independence activities being undertaken in the Texas Coahuila borderlands, where he helped to draft the constitution of the Republic of Texas and served as the first vice-president of the independent republic in 1836.

Like Zavala, Cuban-born elite Loreta Velazquez moved nimbly across shifting national borders, took part in determining the final shape of those boundaries, and wrote extensively about the cultural, political, and geographic theater of war in which she found herself. *The Woman in Battle* (1876) offers a sensationalized memoir of Velazquez's elopement with a Confederate soldier while she was in New Orleans and her subsequent engagement as a cross-dressing Confederate soldier and spy in the U.S. Civil War. Just as her writing challenges many of the precepts governing how we tend to think about the U.S. Civil War—that it was a conflict between U.S. North and South only, of concern primarily to U.S. citizens, and fought by men—so, too, does Torres's description of her travels from Mexico to Galveston, Houston, and El Paso in *Memorias de mi Viaje (Recollections of My Trip)* (1918) challenge many of the precepts governing our understanding of national borders. *Memorias* does not so much chronicle a simple border passage (Torres's dislocation from Mexico City to Galveston because of her father's political persecution by Zapata's forces) as it depicts the complex ethnography of a greater Gulf region system, comprised of different nations' outlying areas. When she lands in Galveston Harbor on July 8, 1914, Torres encounters a circum-Gulf gateway settlement solely comprised of a "wooden shack" and "a dirty wooden bench." It is a first encounter that causes her to ask incredulously "Is this the United States?"[17] Unlike Boston writer Edward Everett Hale's 1886 description of the same border crossing, in which he describes "the picturesque country" of Mexico yielding dramatically as he crosses the border to a vision of "thrift and prosperity on every side" with "prim little citizens of the U.S. walk[ing] the platform in jacket and trousers, boots and straw hats,"[18] Torres's narrative emphasizes this particular borderlands less as a clearly legible national crossroads than as a Gulf circuit in which concepts of nation are blurred and largely secondary.

In "The Future of America," Julian Hawthorne provocatively speculates on the logical end point of such an American nation—a nation characterized by permeable borders, highly mobile subjects and citizens, and a myriad of regional communities. In Hawthorne's estimation, "it is not the descendants of the *Mayflower* who are the representative Americans of the present day," but rather "our foreign-born population," especially those most recently finding themselves within expanding national boundaries. Because Hawthorne—like O'Gorman, Mignolo, and others—recognizes that America is, among other things, "an idea rather than a place: a moral rather than geographical expression," it need not, for Julian Hawthorne at least, be bounded by the nation's geographic borders. In fact, because, according to Hawthorne, "the genuine American spirit" of "enthusiasm for liberty" deteriorates in "direct ratio with the length of an individual's residence in America," the nation should only be led by those "who have not suffered the enervating influence of residency." For him, the ultimate American citizen and patriot is the individual who exists at the interstices between nations and at the periphery rather than the center of the U.S. nation—the man who is "an outlaw in his own country and has never set foot in this."

When taken as a whole, such writings by individual authors like Whitman, Hawthorne, Zavala, Torres, and Velazquez, among others, begin to suggest the limitations of what has historically been a founding premise of American literary studies—that there is a uniform, clearly demarcated, national tradition to which writers contribute and respond, and that such a tradition generates our collective object of critical inquiry. Once we place American writing in the context of the genesis and development of the idea of nation and nationalism in the Americas, the firmly fixed geopolitical parameters that sharply demarcate one nation from another—and one national literary tradition from another—begin to blur. Coming closer to approximating Whitman's 1891 vision of the chorus of voices that might constitute American literature writ large, of necessity, involves reckoning with the history of the very idea of nation and nationalism in the Americas and the place of such key concepts in the writing of authors ranging from Whitman to Velazquez. In other words, the borders of American literature—like the idea of nation itself in the Americas—become blurred, and authors who have historically been central to a U.S. tradition can become peripheral, while those seemingly outside of such a national tradition can suddenly be seen as integral.

Therefore, scholars of American literature have begun to approach U.S. writing as less distinctive from than integrally interested in and bound to the nations and peoples immediately surrounding and intermingling within it. Bruce Harvey's *U.S. National Narratives and the Representation of the Non-European World, 1830–1865*, for example, explores the rich and diverse travel writings of antebellum U.S. popular authors such as Herman Melville, Martin Delany, Maria Cummins, Ephraim G. Squier, and John L. Stephens. As Harvey shows, narratives about non-U.S. America were as important to filling in the temporal contours of the United States as were stories about the goings-on within its borders. Robert Levine explores how the literary nationalism of a host of writers across the long nineteenth century, from Charles Brockden Brown to Frederick Douglass, contested as well as perpetuated the idea of a firmly fixed and immutable nation upon which literary nationalism, at first glance, seems to depend. By reaching beyond the borders of the United States to Haiti in Douglass's case, Levine gives us a more complex and nuanced picture of the shape and texture of U.S. literary nationalism—a picture in which key U.S. writers challenged as well as partook in nationalizing rubrics. Rachel Adams approaches U.S. literature from the vantage point of a North American rather than a nation-based context. Contending that North America offers an intriguing frame for comparative American studies, Adams investigates how our understanding of key themes, genres, and periods within U.S. cultural study is deepened, and in some cases transformed, when Canada and Mexico enter the picture. Other scholars have taken on the transnational geopolitical groundings of those who write within the borders of the United States—those who have been canonized as the founders of American literature, as well as those who have been overlooked as part of that literary undertaking. Anna Brickhouse, for example, has fundamentally reassessed the literary history of the nineteenth-century United States within trans-American and multilingual contexts. Asserting that 'trans-American" is a more accurate term than "American" to describe the writings of those we term

"American Renaissance" writers such as Hawthorne, James Fenimore Cooper, Herman Melville, and Harriet Beecher Stowe, Brickhouse persuasively illustrates how nineteenth-century American literature, written in English, was inextricably linked with writings and audiences in French and Spanish. In her more recent *The Story of Don Luis de Velasco*, Brickhouse uses the story of the sixteenth-century Native American translator who had crucial interactions with Spanish explorers and colonizers near the area now known as Jamestown to show us a hemispheric Jamestown—a Jamestown narrated not only by Smith but also embedded in the larger history of the Americas. Like Brickhouse, Kirsten Silva Gruesz takes the intersections and overlaps of U.S. and Latin American writing as her explicit focus, analyzing nineteenth-century, Spanish-language print culture in the United States and elucidating the efforts that certain North American writers made to alternately foster and contest a hemispheric consciousness. *Ambassadors of Culture* recognizes the contributions made by Latino poets and journalists to both U.S. literary history and to the construction of a Latino identity. Whereas Rodrigo Lazo uncovers a rich archive of popular filibustering literature written in the United States that narrates and helps to normalize filibustering efforts in Cuba, Laura Lomas explores how the writings and translations that famous Cuban writer and independence leader José Martí undertook during his long residency in New York explain and "translate" U.S. political culture for Cuban audiences. Both Lazo and Lomas show the geographically interwoven and overlaid nature of literary production within the United States. Their work reminds us that U.S. cities like New York and Philadelphia were international publishing venues for Latin American political thinkers and writers, and that famous Latin American authors like Cirilo Villaverde wrote and published his famous nineteenth-century novel of Cuba, *Cecelia Valdes*, while living in Philadelphia. Thus, as these scholars illustrate, to write within the United States is not necessarily to write of the United States, but rather is an undertaking that, of necessity, engages with the area's myriad linkages and connections to disparate parts of the hemisphere.

But they also remind us of the powerful thematic and formal simultaneities that do not necessarily recognize national borders and that can bind writers of different nations in shared literary pursuits. George Handley, for example, has charted how fiction from different nations across the Americas shares a cultural and aesthetic legacy that derives from a hemispheric history of slavery. The commonalities that, as we have seen, are shared by former slave societies in the New World shape postslavery writing across the Americas, and, as Handley shows, major fiction from different nations shares parallel anxieties about genealogy, narrative authority, and racial difference. This writing ultimately yields what Handley terms a New World poetic imagination, articulated most fully by Whitman, Pablo Neruda, and Derek Walcott. Likewise, Monique-Adelle Callahan suggests that the poetry of African diaspora descendants in the American hemisphere opens new spaces of interpretation "between the lines" of national literary traditions and allows us to rethink the nineteenth-century canon according to transatlantic conceptions of slavery and freedom. Her in-depth analysis of three key women poets—Frances Ellen Watkins Harper from the United States, Cristina Ayala from Cuba, and Auta de Souza from Brazil—not only brings to light an important and thus far unrecognized triangular connection between the literatures of the United States, the Caribbean, and Brazil, but also highlights the significance of women poets to the genesis and development of a hemispheric literary tradition. National borders evaporate entirely in the literary analyses of Handley and Callahan, and American literature emerges as a hemispheric phenomenon—derived from experiences of slave and postslave societies throughout the Americas.

I reference this wealth of recent work to suggest the intellectual range and vitality of recent scholarship across the disciplines of literature, history, and cultural studies that has, in various ways, reoriented U.S. culture in hemispheric frameworks. As we have seen, these scholars do not adopt a hemispheric model uniformly—for many, particular regions within the United States operate as hemispheric portals, while for others, questions of the interrelation of the United States as a whole to northern as well as southern regions become key. Yet this work collectively begins with the recognition that the nation-state is only one of a number of possible organizing interpretive frameworks. The hemisphere is not the only rubric within which to pose and begin to answer questions about the movements of peoples across time and space, of course. But the work of the previously referenced scholars attests to the fact that it is a richly suggestive, if not conclusive, frame—not precluding global or, even as Wai Chee Dimock suggests, planetary analyses of cultural formations, but rather adding a key piece to the puzzle of the stories of peoples constellating in a particular part of the world. This focus on a particular region of the globe—that region that has been termed the Western Hemisphere—acknowledges in its very nomenclature that the region was also, of course, invented as much as found, as Arthur Whitaker has importantly documented. Just like the nations that have evolved within the area, the notion of the region as a cohesive entity—a hemisphere—has embedded within it a rich array of cultural material that designates it, among other things, as "Western."

When we consider the intellectual findings of the many scholars who have variously adopted hemispheric approaches, it becomes clear that the most recent wave of research is collectively less in danger of reproducing or reinforcing familiar forms of U.S. dominance than of generating new and unfamiliar interpretive models. When scholars approach their object of inquiry, bearing clearly in mind how nationalism and the idea of nation operate (or do not operate) throughout the hemisphere, their work can tend to highlight the sleights of hand through which particular nations that become dominant over time maintain the perception of their predetermined and natural hegemony. In other words, rather than perpetuating U.S. dominance by appropriating the traditions and writings of other nations for an ever-expanding U.S. cultural corpus, such work refutes the timeless and uncontested nature of clearly defined, transhistorical national borders upon which such an allegation depends.

Recent debates over the utility of adopting hemispheric frameworks can tend to overlook this fact. Much like the often-heated debates over women's studies versus gender studies or critical race theory versus identity-based race studies like African American, Chicano/a, and Native American studies, these recent conversations about the politics implicit in the turn to hemispheric studies can have at their core different investments in identity—in this case, disciplinary as well as geopolitical and ethnonational identity. The resulting friction can be as much a result of the effects as the practice of those engaged in hemispheric approaches to their subject. Just as the turn to questioning rather than assuming gender, race, and sexual differences opened the analytical field to those of differing sexual and racial identities, so, too, does a turn to hemispheric as opposed to nation-based area studies have the potential to diversify the field's practitioners and to open the field imaginary to a broad range of perspectives and practices.

Fully implementing the long-standing mandate initiated by Herbert Bolton and reiterated by Janice Radway that we conceptualize an "American" field imaginary necessitates that we follow our findings to their logical conclusions—that we recognize the importance of national boundaries when it makes intellectual sense, utilize or ignore hemispheric frameworks as our questions dictate, and retool our knowledge base and revisit our intellectual assumptions as necessary. Rather than focusing on the identity-based conflicts that can be generated by hemispheric approaches to the fields of American history, literature, religion, sociology, and cultural study, we are challenged to adjust our methods, reorganize our knowledge systems, and alter our fields as our questions necessitate and as the evidence dictates. When confronted by no surviving written "... records of native ..." Mesoamerican population size before Spanish occupancy, world-renowned Latin American historian Woodrow Borah unorthodoxly turned to records of food production in order to see what had not been before visible—that the native population of Mesoamerica fell from ten to two million people between the sixteenth and mid-seventeenth centuries due to European disease and the imposition of Spanish labor practices. When asked, in the wake of such dramatic findings, to describe what a scholar does, Borah succinctly answered, "[H]e follows the data." We need to "follow the data," as Borah did, and see what new worlds it opens for us.[19]

Notes

1. "Annual address of the president of the American Historical Association, delivered at Toronto, December 28, 1932." *American Historical Review*, Volume 38 (April 1933) 3: 448–474: 448.
2. Janice Radway similarly urged scholars to stop using "America" as a default term for the United States and contemplated renaming the association the Inter-American Studies Association.
3. For an informative overview of the history of hemispheric studies, see Ralph Bauer, "Hemispheric Studies," *PMLA* Jan. 2009, 124.1, 234–243.
4. John Adams to Joseph Hawley, 23 June 1776, in Paul H Smith, ed., *Letters of Delegates to Congress 1774–1789*, 26 vols. (Washington, D.C., 1976–2000), 2:385–386.
5. William Williams to Ezekial Williams, 23 August 1776, in ibid., 25:587.
6. James Madison to Thomas Jefferson, 18 November 1781, in ibid., 18:206.
7. James. M. Varnum to William Greene, 2 April 1781, in ibid., 17:115–117.
8. George Templeton Strong, November 8, 1854, in Allan Nevins and Milton Hasley Thomas, eds., *The Diary of George Templeton Strong* (New York: 1952), Vol. 2, *The Turbulent Fifties, 1850–1859*, 196–197; and Vol. 3, *The Civil War, 1860–1865*, 109.
9. Charles Sumner, "Are We a Nation? Address of Hon. Charles Sumner before the New York Young Men's Republican Union, at the Cooper Institute, Tuesday Evening, Nov. 19, 1867," New York, 1867, 4–5.
10. Thomas Wentworth Higginson, "Americanism in Literature," *Atlantic Monthly* (January 1870): 56–63, 56–57.
11. George O. Bartlett to Mr. Ira Andrews, 7 January 1863, Bartlett Papers, Gilder-Lehrman Collection, Pierpont Morgan Library, New York.
12. William Croswell Dome, "Patriotism: Its Defects, Its Dangers and Its Duties," *North American Review* (March 1898): 316–317.

13. Walt Whitman, "July 20, 1883, Letter to the Tertio Millennial Anniversary Association of the City of Santa Fe," np, Huntington Library.
14. Walt Whitman, "Have We a National Literature?" *North American Review*, March 1891. Proof Sheet, Huntington Library 307504.
15. *Original Manuscript by Julian Hawthorne* (St. Louis: WMK Bixby, 1884), np.
16. Lorenzo de Zavala, *Viaje a los Estados Unidos del Norte de América*, Wallace Woolsey, trans., John-Michael Rivera, ed. (Houston: Arte Público Press, 2005), 1.
17. Olga Beatriz Torres, *Memorias de mi Viaje (Recollections of My Trip)*, Juanita Luna-Lawhn, trans. (Albuquerque: University of New Mexico Press, 1994), 47.
18. Edward Everett Hale and Susan Hale, *A Family Flight Through Mexico* (Boston: Lothrop, Lee and Shepard Co, 1886), 266–267.
19. Because this essay is meant to provide an overview and offer a way of thinking about the benefits and challenges of hemispheric studies, I thought it would be useful to include a list of materials for further study: Adams, Rachel. *Continental Divides: Remapping the Cultures of North America* (University of Chicago Press, 2010); Alegría, Fernando. *Walt Whitman en Hispanoamérica* (Mexico City: Fondo de Cultura Económica, 1954); Bauer, Ralph. *The Cultural Geography of Colonial American Literatures: Empire, Travel, Modernity* (New York: Cambridge University Press, 2003); Bauer, Ralph. "Hemispheric Studies." *PMLA* 124.1 (January 2009): 234–245; Brickhouse, Anna. *Transamerican Literary Relations and the Nineteenth-Century Public Sphere* (Cambridge University Press, 2004); Brickhouse, Anna. *The Story of Don Luis de Velasco* (New York: Oxford University Press, 2010); Bell, Gale Chevigny and Gari Laguardia, eds. *Reinventing the Americas: Comparative Studies of Literature of the United States and Spanish America* (New York: Cambridge University Press, 1986); Callahan, Monique-Adelle. *Between the Lines* (New York: Oxford University Press, forthcoming); Dash, J. Michael. *The Other America: Caribbean Literature in a New World Context* (University of Virginia Press, 1998); Dimock, Wai Chee and Lawrence Buell, eds. *Shades of the Planet: American Literature as World Literature* (Princeton University Press, 2007); Dimock, Wai Chee. *Through Other Continents: American Literature Across Deep Time* (Princeton University Press, 2006); Doyle, Don and Pamplona, Marco Antonio, eds. *Nationalism in the New World* (Athens: University of Georgia Press, 2006); Firmat, Gustavo Pérez. *Do the Americas Have a Common Literature?* (Durham: Duke University Press, 1990); Fitz, Earl. *Rediscovering the New World: Inter-American Literature in a Comparative Context* (University of Iowa Press, 1991); Giles, Paul. "Commentary: Hemispheric Partiality," in "Hemispheric American Literary History," Caroline Levander and Robert Levine, eds. *American Literary History* (Fall 2006) 18.3: 648–656; Gruesz, Kirsten Silva. *Ambassadors of Culture: The Transamerican Origins of Latino Writing* (Princeton University Press, 2002); Guterl, Matthew. *American Mediterranean: Southern Slaveholders in*

the Age of Emancipation (Boston: Harvard University Press, 2008); Handley, George. *Postslavery Literature in the Americas: Family Portraits in Black and White* (University of Virginia Press, 2000); Handley, George. *New World Poetics: Nature and the Adamic Imagination in Whitman, Neruda, and Walcott* (Athens: University of Georgia Press, 2007); Hanke, Lewis, ed. *Do the Americas Have a Common History?: A Critique of the Bolton Theory* (New York: Alfred A. Knopf, 1964); Harvey, Bruce. *American Geographies: U.S. National Narratives and the Representation of the Non-European World, 1830–1865* (California: Stanford University Press, 2001); Langley, Lester. *The Americas in the Age of Revolution: 1750–1850* (New Haven: Yale University Press, 1998); Lazo, Rodrigo. *Writing to Cuba: Filibustering and Cuban Exiles in the United States* (University of North Carolina Press, 2005); Levander, Caroline and Robert Levine, editors. *Hemispheric American Studies* (Rutgers University Press, 2008); Levine, Robert. *Dislocating Race and Nation: Episodes in Nineteenth-Century American Literary Nationalism* (University of North Carolina Press, 2008); McClennen, Sophia. "Inter-American Studies or Imperial American Studies?" *Comparative American Studies* 3.4 (2005): 393–413; Mignolo, Walter. *The Darker Side of the Renaissance: Literacy, Territoriality, & Colonization* (University of Michigan Press, 2003); Morieras, Alberto. *The Exhaustion of Difference: The Politics of Latin American Cultural Studies* (Durham: Duke University Press, 2001); Rojo, Antonio Benítez. James Maraniss, trans. *The Repeating Island: The Caribbean and the Postmodern Perspective* (Duke University Press, 1997); Saldívar, José David. *The Dialectics of Our America: Genealogy, Cultural Critique, and Literary History* (Duke University Press, 1991); Scott, Rebecca. *Degrees of Freedom: Louisiana and Cuba after Slavery* (Harvard University Press, 2008); Shukla, Sandya and Heidi Tinsman, eds. *Imagining Our Americas: Toward a Transnational Frame* (Durham: Duke University Press, 2007); Sommer, Doris. *Proceed with Caution, When Engaged by Minority Writing in the Americas* (Harvard University Press, 1999); Sommer, Doris. "Supplying Demand: Walt Whitman and the Liberal Self," in *Reinventing the Americas: Comparative Studies of Literature of the United States and Spanish America*, Eds. Bell Gale Chevigny and Gari Laguardia (Cambridge: Cambridge University Press, 1986): 68–91; Spillers, Hortense. *Comparative American Identities: Race, Sex and Nationality in the Modern Text* (New York: Routledge, 1991); Whitaker, Arthur. *The Western Hemisphere Idea: Its Rise and Decline* (Ithaca: Cornell University Press, 1954).

CHAPTER 3

Coloniality at Large: The Western Hemisphere and the Colonial Horizon of Modernity

Walter D. Mignolo
(Translated by Michael Ennis)

Before the Cold War, the closest the United States had ever come to a permanent foreign policy was in our relationship with the nations of the Western Hemisphere. In 1823, the Monroe Doctrine proclaimed our determination to insulate the Western Hemisphere from the contests over the European balance of power, by force if necessary. And for nearly a century afterward, the causes of America's wars were to be found in the Western Hemisphere: in the wars against Mexico and Spain, and in threats to use force to end Napoleon III's effort to install a European dynasty in Mexico.[1]

The thesis that I propose and defend here is that the emergence of the idea of the "Western Hemisphere" gave way to a radical change in the imaginary and power structures of the modern/colonial world.[2] This change not only had an enormous impact in restructuring the modern/colonial world, but it also had—and continues to have—important repercussions for South-North relations in the Americas, for the current configuration of "Latinidad" in the United States, as well as for the diverse Afro-American communities in the North, South, and Caribbean.

I use the concept of "imaginary" in the sense in which the Martinican intellectual and writer Eduardo Glissant uses it.[3] For Glissant, "the imaginary" is the symbolic world through which a community (racial, national, imperial, sexual, etc.) defines itself. In Glissant, the term has neither the common meaning of a mental image, nor the more technical meaning that it has in contemporary psychoanalytic discourses, in which the imaginary forms a structure of differentiation between the symbolic and the real. Departing from Glissant, I give the term a geopolitical meaning and use it in terms of the foundation and formation of the imaginary of the modern/colonial world system. The image of Western civilization that we have today is the result of the long process of constructing the "interior" of that imaginary, from the transition of the Mediterranean as center to the formation of the Atlantic commercial circuit, just as it constructed its "exteriority." In the West, men and women of letters, travelers, statesmen of every kind, ecclesiastical functionaries, and Christian thinkers have constructed the "interior" image. This was always accompanied by an "internal exterior," which is to say, by an "exteriority" but not by an "exterior." Until the end of the fifteenth century, European Christianity was on the margins of the world system, and had identified itself with Japhet and the West, distinguishing itself from Asia and Africa. This Occident of Japhet was also the Europe of Greek mythology. From the beginning of the sixteenth century, with the triple concurrence of the defeat of the Moors, the expulsion of the Jews, and European expansion across the Atlantic, Moors, Jews, and Amerindians (and, with time, African slaves as well) all became configured in the Western, Christian imaginary as the difference (exteriority) in the interior of the imaginary. Toward the end of the sixteenth century, the Jesuit missions in China added a new dimension of "exteriority"—the outside that is within precisely because it contributed to the definition of itself.

The Jesuits contributed in the extremes (Asia and America) to constructing the imaginary of the Atlantic commercial circuit that, with various historical versions, came to shape the contemporary image of Western civilization, which I will return to later in the paper. However, the imaginary about which I am speaking is not only constituted in and by colonial discourse, including colonial discourse's own internal differences (e.g., Las Casas and Sepúlveda; or the discourse from northern Europe, which from the end of the seventeenth century drew a border between itself and southern Europe, thus establishing the imperial difference), but it is also constituted by the responses (or, in certain moments, the lack of responses) of the communities (empires, religions, civilizations) that the Western imaginary involved in its own self-description. Although these features are planetary, in this article, I will limit myself to examining the responses from the Americas to the discourse and integrated politics that in different moments differentiated Europe first, then the Western Hemisphere, and, finally, the North Atlantic.

One might ask, "What do I mean by modern/colonial world or modern/colonial world system?" I take as my point of departure the metaphor of the modern world system proposed by Wallerstein.[4] The metaphor has the advantage of marking a historical and relational framework for reflections that escape the national ideologies under which continental and subcontinental imaginaries were forged, as much in Europe as in the Americas, over the last 200 years. I am not interested in determining how old the world system is, whether it is 500 or 5,000 years old.[5] I am even less interested in knowing the age of modernity or capitalism.[6] What interests me is the emergence of the Atlantic commercial circuit in the sixteenth century, which I consider fundamental to the history of capitalism and modernity/coloniality. I am not interested in arguing about whether or not there was commerce prior to the emergence of the Atlantic commercial circuit before the sixteenth century. Rather, I am interested in the impact that the emergence of the Atlantic commercial circuit had on the formation of the modern/colonial world in which we are living and bearing witness, and on the global transformations that accompanied that moment. Although I take the idea of the world system as my point of departure, I stray from it to introduce the concept of "coloniality" as the other side (the darker side?) of modernity. By using "coloniality," I do not mean to say that the metaphor of the world system has not considered colonialism. On the contrary, what I assert is that the metaphor of the modern world system leaves in darkness the coloniality of power and the colonial difference.[7] Consequently, the modern world system is only conceived from its own imaginary, and not from the conflictive imaginary that rises up with and from the colonial difference. Indigenous rebellions and Amerindian intellectual production from the sixteenth century on, just like the Haitian Revolution at the beginning of the nineteenth century, are constitutive moments of the imaginary of the modern/colonial world and not mere occurrences in a world constructed from Hispanic discourses (for example, the Sepúlveda/Las Casas debate about the "nature" of the Amerindian, in which the Amerindian had no place to give his or her opinion; or the French Revolution, which is considered by Wallerstein the foundational moment of the geoculture of the modern world system).[8] In this sense, the contribution of Anibal Quijano in an article co-written with Wallerstein is a fundamental theoretical turn in outlining the conditions under which the coloniality of power was and is a strategy of "modernity," from the moment of the expansion of Christianity beyond the Mediterranean (America, Asia), which contributed to the self-definition of Europe, and has been indissociable from capitalism since the sixteenth century.⁹ This moment in the construction of the colonial imaginary, which later will be taken up and transformed by England and France in the project of the "civilizing mission," does not appear in the history of capitalism given by Arrighi.¹⁰ In Arrighi's reconstruction, the history of capitalism is seen either from "within" (in Europe) or from within toward the outside (from Europe toward the colonies). Therefore, from Arrighi's perspective, the coloniality of power is invisible. Consequently, capitalism, like modernity, appears as a European phenomenon and not a global one, in which all the world participates, albeit with distinct positions of power. That is, the coloniality of power is the axis that organized and organizes the colonial difference—the periphery as nature.

Under this general panorama, I am interested in recalling a paragraph by Quijano and Wallerstein that offers a framework through which to understand the importance of the idea of the "Western Hemisphere" in the imaginary of the modern/colonial world since the beginning of the nineteenth century:

> The modern world-system was born in the long sixteenth century. The Americas as a geo-social construct were born in the long sixteenth century. The creation of the geo-social entity, the Americas, was the constitutive act of the modern world system. The Americas were not incorporated into an already capitalist world-economy. There could not have been a capitalist world-economy without the Americas.¹¹

Leaving aside the particularistic and triumphalistic connotations that the paragraph could invoke, as well as a discussion of whether or not there would have been a global capitalist economy without the riches of American mines and plantations, the fact is that the capitalist economy changed course and accelerated with the emergence of the Atlantic commercial circuit. The transformation of the Aristotelian conception of slavery was required as much by the new historical conditions as by the human type (e.g., Negro, African) that was identified from the beginning of that moment with slavery and established new relations between race and labor. Starting from this moment, the moment of the emergence and consolidation of the Atlantic commercial circuit, it was already impossible to conceive of modernity without coloniality, the side silenced by the reflexive image that modernity (e.g., the intellectuals, official state discourses) constructed of itself and that postmodern discourse critiques, from the interiority of modernity, as a self-image of power. Postmodernism, self-conceived in the unilateral line of the history of the modern world, continues to obscure coloniality and maintains a universal and monotopical logic—from the left as well as the right—from Europe (or the North Atlantic) toward the outside. The colonial difference (imagined in the pagan, the barbarian, the underdeveloped) is a passive place in postmodern discourses. What postmodernism does not want to say is that it is in reality a passive place in modernity and in capitalism. The visibility of the colonial difference in the modern world began to be noted with the decolonization (or independence) movements from the end of the eighteenth century until the second half of the twentieth century. The emergence of the idea of the "Western Hemisphere" was one of those movements.

However, we should remember that the emergence of the Atlantic commercial circuit had the particularity (and this aspect is important for the idea of the "Western Hemisphere") of connecting the commercial circuits already existent in Asia, Africa, and Europe (the commercial network in which Europe was the most marginal space to the center—China with Anahuac and Tawantinsuyu, the two great circuits, disconnected until then from the aforementioned circuits, as much by the Pacific as by the Atlantic).[12]

The imaginary of the modern/colonial world is not the same when viewed from the history of ideas of Europe as when looked at from the perspective of colonial difference: the histories forged by the coloniality of power in the Americas, Asia, or Africa. These are the histories of the cosmologies prior to contact with Europe since the sixteenth century, just as in the constitution of the modern colonial world, the states and societies of Africa, Asia, and the Americas had to respond—and did respond—in different ways and at different historical moments. Europe, by way of Spain, took the sword to North Africa and Islam in the sixteenth century; China and Japan were never under Western imperial control, although they could not avoid responding to its expansionary efforts, above all since the nineteenth century, when Islam renewed its relations with Europe.[13] South Asia, India, and several sub-Saharan African nations were the objectives of emergent colonial powers—England, France, Belgium, and Germany. The configuration of modernity in Europe and coloniality in the rest of the world (with exceptions, to be sure, as is the case in Ireland) was the hegemonic image sustained in the coloniality of power that makes it difficult to think that modernity could have existed without coloniality. Indeed, coloniality is constitutive of modernity, not derivative of it.

The Americas, above all in the early experiences of the Caribbean, Mesoamerica, and the Andes, established the model for the imaginary of the Atlantic circuit. Beginning with this moment, we find transformations and adaptations of the model of colonization and of the religious-epistemological principles that were imposed from then on. Numerous examples can be invoked here, beginning in the sixteenth century, and fundamentally in the Andes and Mesoamerica.[14] I prefer, however, to summon more recent examples, in which modernity/coloniality persists in its dual aspect. Indeed, the imaginary persists as much in its hegemonic imaginary, despite its transformations, as in the constant adaptations from the planetary colonial exteriority. This is an exteriority that is not necessarily outside of the West (which would mean a total lack of contact), but which is an interior exteriority and exterior exteriority (the forms of resistance and opposition trace the interior exteriority of the system). This duality fits very well, for example, with the way in which the Spanish state, as well as different American states, celebrated the 500-year anniversary of the discovery of America in the face of indigenous movements and intellectuals that protested the celebration, attempting to re-inscribe the history of the conquest. The Laguna novelist Leslie Marmon Silko included a "map of the five hundred years" in her novel *Almanac of the Dead*, published one year before the quincentennial.[15]

The first declaration from the Lacandon Forest, in 1994, began by saying, "We are the product of 500 years of struggle." Rigoberta Menchú, in a report read at the conference organized by sociologist Pablo González Cassanova on democracy and the multiethnic state in Latin America, also evoked the marker of 500 years of oppression:

> The history of the Guatemalan people can be interpreted as a concentration of the diversity of America, of the chosen fight, forged from the bases and in many parts of America, still maintained in forgetfulness. Forgetfulness not because it is wanted, but because a tradition in a culture of oppression has returned. Forgetfulness that requires a fight and a resistance by our peoples that has a 500-year history.[16]

Thus, this frame of 500 years is the frame of the modern/colonial world from distinct perspectives of its imaginary, which does not reduce the confrontation between the Spanish and Amerindians, but extends it to Creoles (white, black, and mestizo) springing from the importation of African slaves, whom the white European population transplanted in their own interest in the majority of cases to the Americas. That ethnoraciality is the point of articulation of the imaginary, constructed in and beginning with the Atlantic commercial circuit, does not exclude aspects of class (which were given entrance in the distributions and transformations that slavery suffered), as was known in the Mediterranean beginning in 1517, when the first 15,000 slaves were transported from Africa. Nor does it deny the aspects of gender and sexuality that Tressler analyzed recently. I mean to say only that the ethnoraciality became the machinery of colonial difference. Beginning with the expulsion of the Moors and the Jews, it was configured from the debates over the place of the Amerindians in the economy of Christianity, and, finally, by the exploitation and silencing of African slaves. It was with and from the Atlantic commercial circuit that slavery became synonymous with blackness.

This view is not a description of colonialism, but of coloniality, of the construction of the modern world in the exercise of the coloniality of power. It is also a description of the responses from the colonial difference to the programmed coercion that the coloniality of power exercises. The imaginary of the modern/colonial world arose from the complex articulation of forces, of voices heard or silenced, of memories compact or fractured, of histories told from only one side that suppress other memories, and of histories that were and are told from the double consciousness that generates the colonial difference. In the sixteenth century, Sepúlveda and Las Casas contributed, in different ways and from different political positions, to the construction of colonial difference. Guaman Poma, or Ixtlixochitl, thought and wrote from the colonial difference in what was situated by the coloniality of power. At the beginning of the twentieth century, the sociologist and black intellectual W. E. B. Du Bois introduced the concept of "double consciousness," which captures the dilemma of subjectivities formed within the colonial difference: that is, the experiences of anyone who lived and lives modernity

> from coloniality. This is a strange sensation in this America, says Du Bois, for anyone who does not have a true self-consciousness but whose consciousness must form itself and define itself with relation to the "other world."[17] That is, the consciousness lived from the colonial difference is double because it is subaltern. Colonial subalternity generates diverse double consciousnesses, not only African American, which is Du Bois's experience, but also the "consciousness that gave birth to Rigoberta Menchú" or "the consciousness of the new mestiza" in Gloria Anzaldúa.[18]

Let us cite Du Bois:

> It is a particular sensation, this double-consciousness, this sense of always looking at one's self through the eyes of the others, of measuring one's

soul by the tape of a world that looks on in amused contempt and pity. One ever feels his two-ness—an American, a Negro; two souls, two thoughts, two unreconciled strivings; two warring ideals in one dark body.... The history of the American Negro is the history of this strife—this longing to attain self-conscious manhood, to merge his double self into a better and truer self.[19]

The beginning of double consciousness is, in my argument, the characteristic of the imaginary of the modern/colonial world from the margins of the empires (the Americas, Southeast Asia, North and Sub-Saharan Africa, Oceania). Double consciousness, in sum, is the consequence of the coloniality of power and the manifestation of subjectivities forged in the colonial difference. The local histories vary because the history of Europe was changing in the process of forming itself in the expansive movement of the West. In the continental and subcontinental divisions established by Christian symbolic cartography (e.g., the continental trilogy of the known world at that time: Europe, Africa, and Asia), the colonial horizon of the Americas is foundational to the imaginary of the modern world. The emergence of the "Western Hemisphere" as an idea was a moment of the imaginary arising in and with the Atlantic commercial circuit. The particularity of the image of the "Western Hemisphere" marked the insertion of the Creole descendents of Europeans, in both Americas, into the modern/colonial world.

This insertion was, at the same time, the consolidation of the Creole double consciousness that was forging itself in the same process of colonization.

Creole Double Consciousness and the Western Hemisphere

The idea of the "Western Hemisphere" (which only appears as such in cartography at the end of the eighteenth century) establishes an ambiguous position. America simultaneously constitutes difference and sameness. It is the other hemisphere, but it is Western. It is distinct from Europe (of course, it is not the Orient), but it is bound to Europe. It is different, however, from Asia and Africa, continents and cultures that do not form part of the Western Hemisphere. But who defines such a hemisphere? For whom is it important and necessary to define a place of possession and difference? Who experienced the colonial difference as Creoles of Hispanic (Bolívar) and Anglo-Saxon (Jefferson) descent?

As we might expect, what each one understood by "Western Hemisphere" (although the expression originated in the English part of the Americas) differs, and does so in a manner that is far from trivial. In the "Carta de Jamaica," which Bolívar wrote in 1815 and sent to Henry Cullen, "a gentleman of this island," the enemy was Spain. Bolívar's references to "Europe" (the north of Spain) were not references to an enemy, but the expression of a certain surprise before the fact that "Europe" (which supposedly Bolívar would locate at that time in France, England, and Germany) would show itself to be indifferent to the struggles for independence that were occurring during those years in Hispanic America. Considering that England was already a developing empire with several decades of colonization experience in India and the enemy of Spain, it is possible that Mr. Cullen received Bolívar's diatribes against the Spanish with interest and pleasure. The "black legend" remained a trademark in the imaginary of the modern/colonial world.

On the other hand, Jefferson's enemy was England, although, contrary to Bolívar, Jefferson did not reflect on the fact that Spain was not incensed by the independence of the United States of North America. With this I wish to say that the crossed references—Jefferson toward the south and Bolívar toward the north—really were crossed references. While Bolívar imagined, in his letter to Cullen, the possible political organization of America (which in his imaginary was Hispanic America) and speculated starting from the suggestions of a dubious writer of dubious stock, Abe de Pradt, Jefferson looked with enthusiasm on the independence movements in the South, although he was suspicious of the path of their political future.[20] In a letter to Baron Alexander von Humboldt, dated December 1813, Jefferson thanked him for sending astronomical observations after the journey that Humboldt had made through South America and emphasized the opportunity of the trip in the moment when "those countries" were in the process of "becoming actors on their stage," adding:

> That they will throw off their European dependence I have no doubt; but in what kind of government their revolution will end I am not so certain. History, I believe, furnishes no example of a priest-ridden people maintaining a free, civil government...but in whatever governments they end they will be "American" governments, no longer to be involved in the never-ceasing broils of Europe.[21]

For his part, Bolívar expressed vehemently:

> I want more than anything to see the formation in America of the greatest nation in the world, less for its extension and riches than for its liberty and glory. Although I aspire to the perfection of the government of

my homeland, I cannot persuade myself that the New World is for the moment governed by a great republic.²²

While Bolívar writes of the "hemisphere of Columbus," Jefferson spoke of the hemisphere that "America has for itself." In reality, Bolívar and Jefferson thought about two Americas. And they were different geographically too. The Iberian America extended to what are today California and Colorado, while Anglo-Saxon America did not go further west than Pennsylvania, Washington, and Atlanta.

Where Bolivar and Jefferson met was in the way they referred to their respective metropols, Spain and England. Referring to the conquest, Bolívar underscored the "barbarities of the Spanish" as "barbarities that the present age has rejected as fabulous, because they seem beyond human perversity."²³ Jefferson refers to the English as exterminators of the Native Americans ("extermination of this race in *our* America," emphasis added, WM), as another chapter "in the English history of the same colored man in Asia, and of the brethren of their own color in Ireland, and wherever else Anglo-mercantile cupidity can find a twopenny interest in deluging the earth in human blood."²⁴ Even though the references were crossed, there was this in common with Bolívar and Jefferson: The idea of the Western Hemisphere was linked to the rising of Anglo and Hispanic Creole consciousness. The emergence of black Creole consciousness in Haiti was different because it was limited to French colonialism and the African heritage. French colonialism, like English colonialism in the Caribbean, did not have the force of English immigration that was the foundation of the United States. Nor did French colonialism have the legacies of the strong Hispanic colonialism. Black Creole consciousness, contrary to white Creole consciousness, was not inherited from colonizers and emigrants. Rather, it was inherited from slavery: The idea of a "Western Hemisphere," or, as Martí would say later, "our America," was not common among black Creoles. In sum, "Western Hemisphere" and "our America" are fundamental figures of the Creole imaginary, Saxon and Iberian, but not of the Amerindian imaginary (in the North and in the South) or the Afro-American imaginary (as much in Latin America as in the Caribbean and North America). We know, for example, what Jefferson thought of the Haitian Revolution and "that race of men."²⁵ Creole consciousness in relation to Europe was forged more as a geopolitical consciousness than a racial one. However, Creole consciousness as a racial consciousness was forged internally in the difference with Amerindian and Afro-American populations. The colonial difference was transformed and reproduced during the national period, and it is that transformation that has been termed "internal colonialism." Internal colonialism is, then, the colonial difference exercised by the leaders of national construction. This aspect of the formation of white Creole consciousness is what transformed the imaginary of the modern/colonial world system and established the basis for internal colonialism that crosses every period of national formation, as much in Iberian America as in Anglo-Saxon America.[26] The ideas of "America" and the "Western Hemisphere" (not the "West Indies," which was a Hispanic designation for colonial territories) were imagined as places of possession and the right to self-determination. Although Bolívar thought of his nation as belonging in the rest of America (Hispanic), Jefferson thought about something more indeterminate, although what he thought was the memory of Saxon colonial territoriality, a territory that had not been configured by the idea of "the West Indies." "The West Indies" was the distinct mark of Hispanic colonialism that differentiated its possessions in America from those in Asia (e.g., the Philippines), which were identified as the "East Indies." In the formation of New England, on the other hand, "West Indies" was a foreign concept. When the expression was introduced into English, "West Indies" was fundamentally used to designate the English Caribbean. What was clear for both Bolívar and Jefferson was the geopolitical separation from Europe, a Europe that in one case had its center in Spain, and in the other case, in England. Since the previous designations (West Indies, America) were formulated in the Spanish and European consciousness, "Western Hemisphere" was the necessary, distinctive sign for the imaginary of postindependence white Creole consciousness. The Creole consciousness was not, to be sure, a new event, since there would not have been independence, in the North or South, without it. What was new and important in Jefferson and Bolívar was the moment of transformation of the colonial Creole consciousness into a postcolonial and national Creole consciousness, and the emergence of internal colonialism against the Amerindian and Afro-American populations.

From the perspective of black Creole consciousness, as Du Bois describes, we can say that the white Creole consciousness is a double consciousness that was not recognized as such. The denial of Europe was not, either in Hispanic America or in Anglo-Saxon America, the denial of Europeanness, since in both cases, and in every impulse of white Creole consciousness, they tried to be American without ceasing to be European by being Americans who were still different from Amerindians and Afro-Americans.

If in geopolitical terms, Creole consciousness was defined with respect to Europe, in racial terms, it was defined with respect to black Creoles and Amerindian peoples. Creole consciousness lived (and still lives) as double, although it did not and does not recognize itself as such. It was recognized instead in the homogeneity of the national imaginary and, from the beginning of the twentieth century, in mestizaje as the contradictory expression of homogeneity. The celebration of the pure mestizaje by blood says it all. The formation of the nation-state required homogeneity more than dissolution; therefore, the celebration of heterogeneity was unthinkable, or, better, heterogeneity had to be hidden. If it had not been thus, if the white Creole consciousness had recognized itself as double, we would not have the problems of identity, multiculturalism, and pluriculturality that we have in the United States, Hispanic America, and the Caribbean. Jefferson wrote:

> The European nations constitute a separate division of the globe; their localities make them part of a distinct system; they have a set of interests of their own in which it is our business to never engage ourselves. America has a hemisphere to itself.[27]

Jefferson denies Europe, not Europeanness. The Haitian revolutionaries Toussaint L'Ouverture and Jean-Jacques Dessalines, on the other hand, deny Europe and Europeanness.[28] Directly or indirectly, it was the African Diaspora and not the Western Hemisphere that fed the imaginary of the Haitian revolutionaries. On the other hand, the vehemence with which Bolívar and Jefferson proposed the separation from Europe was motivated by knowing themselves and feeling themselves to be, in the last instance, Europeans on the margin, Europeans who were not Europeans, but who wanted to be so at their very core. This white Creole double consciousness, varying in intensity in the colonial and national periods, was the sign and the legacy of the independent intellectuality of nineteenth-century national consciousness. I repeat that the characteristic of this double consciousness was not racial but geopolitical and defined itself in relation to Europe. The double consciousness was not manifested, to be sure, in relation to Amerindian or Afro-American components of the population. From the Creole point of view, how to be Creole and Indian or black at the same time was not a problem that had to be resolved. In this context—in relation to Amerindian and Afro-American communities—white Creole consciousness defined itself as homogeneous and different. If the white Creoles did not realize what their double consciousness was due to, I suggest that one of the traits of the conceptualization of the Western Hemisphere was the integration of America into the West, which was not possible for black Creole consciousness: Africa, because of its geographic localization, never was part of the Western geopolitical imaginary. Du Bois was not permitted, like Guaman Poma de Ayala or Garcilaso de la Vega in the sixteenth century, to feel himself part of Europe or as some form of European on the margin. Varied forms of double consciousness, finally, were the consequences and are the legacies of the modern/colonial world.

The Western Hemisphere and the Geoculture of the Modern/Colonial World System

One of the traits that distinguishes the processes of decolonization in the Americas in the end of the eighteenth century and the beginning of the nineteenth century is, as has been noted by Klor de Alva, the fact that decolonization was in the hands of "Creoles" rather than "natives," as happened in twentieth-century Africa and Asia.[29] There is, however, another important element to keep in mind: The first wave of decolonization was accompanied by the idea of the "Western Hemisphere" and the transformation of the imaginary of the modern/colonial world, which boils down to this geopolitical image.

If the idea of the "Western Hemisphere" found its moment of emergence in the independence of Creoles in both Americas, its moment of consolidation can be found almost a century later, after the Spanish-American War and during the presidency of Theodore Roosevelt, at the dawn of the twentieth century. If histories need a beginning, then the history of the strong rearticulation of the idea of the Western Hemisphere in the twentieth century had its beginning in Venezuela, when armed forces from Germany and England initiated a blockade to pressure for the payment of foreign debts. The Spanish-American War (1898) had been a war for the control of the seas and the Panama Canal against the threats of the well-established imperial nations of Western Europe, a danger that was repeated with the blockade against Venezuela. The intervention of Germany and England was a good moment to revive the call for autonomy for the "Western Hemisphere," which had lost strength in the years prior to and during the American Civil War. The fact that the blockade was against Venezuela created the conditions for the idea and ideology of the "Western Hemisphere" to be revived as not only a question of U.S. jurisdiction, but also of the jurisdiction of Latin American countries. The Argentinean Luís María Drago, Minister of Foreign Affairs, made the first step in that direction in December 1902.[30]

Whitaker proposes in a broad outline an interpretation of these years of international politics that helps us to understand the radical change in the imaginary of the modern/colonial world system that took place at the beginning of the twentieth century with the Rooseveltean reinterpretation of the idea of the "Western Hemisphere." According to Whitaker, Luís María Drago's proposed resolution to the embargo on Venezuela (now known as the "Drago Doctrine") was in reality a sort of "corollary" to the Monroe Doctrine from a multilateral perspective that involved, of course, all of the states of the Americas. Whitaker suggests that Drago's position was not well received in Washington, D.C. because, among other things, the United States considered the Monroe Doctrine a doctrine of national politics and, indirectly, unilateral when applied to international relations. Contrary to U.S. views on the Monroe Doctrine, Drago interpreted it as a multilateral principle valid for the whole Western Hemisphere that could be executed in and from any part of the Americas. The second reason that Washington shunned the Drago Doctrine, according to Whitaker, was a consequence of the first: If, in fact, a corollary had been necessary to extend the effectivity of the Monroe Doctrine to international relations, this "corollary" should have come from Washington, D.C. and not Argentina, or any part of Latin America, for that matter. This was, according to Whitaker, the road Washington, D.C. followed when, in December 1904, Roosevelt proposed his own "corollary" to the Monroe Doctrine. Although similar to Drago's proposal, Roosevelt's had important differences. Whitaker enumerates the following points of similarity:

> (a) both "corollaries" were designed to solve the same problem (European intervention in the Americas) and were based on the same premises (the Monroe Doctrine and the idea of the Western Hemisphere); (b) both "corollaries" proposed to solve the problem through an exception to international law in favor of promoting the Western Hemisphere; and (c) both proposed to achieve this solution through an "American policy pronouncement, not through a universally agreed amendment to international law."[31]

The differences, however, were what reoriented the configuration of the new world order: the "ascent" of one neocolonial or postcolonial country to the group of imperial nation-states—a change of no small measure in the imaginary and structure of the modern/colonial world. The differences between Roosevelt and Drago, according to Whitaker, are found in the manner of implementing the new international politics.

Roosevelt proposed to do it unilaterally, from the United States, while Drago proposed a multilateral action, which would be democratic and inter-American. The results of Roosevelt's "corollary" are very different from what could be imagined to have happened if the Drago Doctrine had been implemented. However, Roosevelt claimed for America the monopoly of rights of the administration of autonomy and democracy in the Western Hemisphere.[32] The Monroe Doctrine, rearticulated with the idea of the "Western Hemisphere," introduced a fundamental change in the configuration of the modern/colonial world and the imaginary of modernity/coloniality. Whitaker's conclusion on this chapter of the modern/colonial world is apt: "As a result [of the implementation of the "Roosevelt corollary" instead of the "Drago corollary"] the leaders in Washington and those in Western Europe came to understand each other better and better as time went on. The same development, however, widened the already considerable gap between Anglo-Saxon America and Latin America."[33]

The moment I have just narrated, based on Whitaker's work, suggesting the connections between international politics and the imaginary of the modern colonial world, appears in the history of Latin American literature as "La Oda a Roosevelt" by the Nicaraguan poet and cosmopolitan Rubén Darío, as well as in the essay "Ariel" by the Uruguayan intellectual Enrique Rodó. I am interested here in returning to the period that extends from the Spanish-American War until the "triumph" of the "Roosevelt corollary" in order to reflect on geoculture and the imaginary of the modern/colonial world, as well as the impact the idea of the Western Hemisphere had on that imaginary.

Responding to criticisms directed at the strong economic aspect of the concept of the modern world system, Immanuel Wallerstein introduced the concept of geoculture.[34] Wallerstein constructs the concept, historically, from the French Revolution until the crisis of 1968 in France, and, logically, as the cultural structure that geoculturally binds to the world system. The "geoculture" of the modern world system should be understood as the ideological (and hegemonic) image sustained and expanded by the dominant class after the French Revolution. The hegemonic image is not equivalent to social structure, but rather the manner in which one group, which imposes the image, conceives social structure. The "imaginary of the modern/colonial world" should be understood as the various and conflicting economic, political, social, and religious perspectives through which social structure is actualized and transformed. But Wallerstein conceives geoculture only in its monotopic and hegemonic aspect, localized in the second modernity, which saw the ascent of France, England, and Germany as leaders of the modern/colonial world.[35] Without doubt, what Wallerstein calls geoculture is the component of the imaginary of the modern/colonial world that universalizes itself, and does so not only in the name of the civilizing mission to the non-European world, but which relegates the sixteenth century to the past, and with it, Southern Europe. The imaginary that emerges with the Atlantic commercial circuit, which puts Iberians, Amerindians, and African slaves into conflictive relations, is not a component of geoculture for Wallerstein. That is to say, Wallerstein describes only the hegemonic imaginary of the modern world system as geoculture, leaving to one side as many contributions from the colonial difference as from the imperial difference (i.e., the emergence of the Western Hemisphere in the colonial horizon of modernity). Wallerstein's geoculture is, then, the hegemonic imaginary of the second phase of modernity; consequently, it is Eurocentric in the strict sense of the word, centered in France, Germany, and England from the perspective of history (from the French national imaginary). The French Revolution takes place precisely at the moment of "inter-imperium," in which the Europe of nations was consolidated by the colonial question. The independence of the United States, which not only anticipated but also contributed to making the French Revolution possible, is other or marginal to Wallerstein's concept of geoculture because, in my interpretation, his concept of the modern world system is blind to colonial difference. This is crucial, because independence in the Americas, the first antisystematic movements, were movements from the colonial difference. These movements were generated by and in the colonial difference, although colonial difference is reproduced through them in different ways, as I mentioned earlier. In the concept of "geoculture," Wallerstein underlines the hegemonic component of the modern world that accompanied the bourgeois revolution in the consolidation of the Europe of nations and that, at the same time, relegated as "peripheral" events that represent the first decolonization movements of a modern, but also colonial, world. Such blindness was notable in the case of the Haitian Revolution, as Trouillot demonstrates.[36] Trouillot explains why a revolution of black Creoles, supported by black slaves, did not have a place in the liberal discourses about the rights of man and citizen, which had been thought in a world where the "invisible matrix" was white, that is, composed fundamentally of white citizens and not Indians or Negroes. In this scheme, the differences of gender and sexuality were subsumed by racial classifications. It was not, nor is it, the same to be a white woman as it is to be a woman of color. Coloniality is constitutive of modernity. Asymmetric relations of power at the same time as the active participation from the colonial difference in the expansion of the Atlantic commercial circuit across the centuries are what justify and make necessary the concepts of "coloniality of power" and "colonial difference" in order to correct the historicogeographic limitations at the same time as the logics of the concept of geoculture in its formulation by Wallerstein:[37]

In the case of the modern world system, it seems to me that its geoculture emerged with the French Revolution and then began to lose its widespread acceptance with the world revolution of 1968. The capitalist world-economy had been operating since the long sixteenth century. *It functioned for three centuries, however, without any firmly established geoculture.* That is to say, from the sixteenth to the eighteenth century, no one set of values and basic rules prevailed within the capitalist world-economy, actively endorsed by the majority of the cadres, and passively accepted by the majority of the ordinary people. The French Revolution, *lato senso*, changed that. It established two principles: (1) the normality of political change and (2) the sovereignty of people... The key point to note about these two principles is that they were, in and of themselves, quite revolutionary in their implications for the world system. Far from ensuring the legitimacy of the capitalist world-economy, they threatened to delegitimize it in the long run. It is in this sense that I have argued elsewhere that the French Revolution represented the first of the antisystemic revolutions of the capitalist world-economy—in a small part, a success; in a larger part, a failure.[38]

Wallerstein's difficulty in recognizing the constitution of the imaginary of the modern/colonial world without the participation of France or England, and therefore, denying the contribution of three centuries of Spanish and Portuguese power, is, without doubt, a consequence of how he conceives geoculture. The Northern European imaginary, beginning with the French Revolution, is the imaginary that was constructed parallel to the triumph of England and France over Spain and Portugal as new imperial powers. The emergence of the concept of the "Western Hemisphere" did not allow foreseeing that it marked, from the beginning, the limits of what Wallerstein calls geoculture. And it marked it in two ways: by articulating the colonial difference and by absorbing, for the length of its history, the concept of the "civilizing mission." Wallerstein places the concept of the "civilizing mission" as central to geoculture; however, the civilizing mission remains a translation of the "christianizing mission," dominant from the sixteenth until the eighteenth century, which Wallerstein does not recognize as geoculture.

Samuel Huntington described the new world order after the end of the Cold War in nine civilizations: the West, Latin America, Africa (more specifically, sub-Saharan Africa), Islam, China, Hindu, Orthodox, Buddhist, and Japanese. Leaving aside the fact that Huntington's classificatory logic seems like the famous Chinese emperor mentioned by Jorge Luis Borges and adopted by Michel Foucault at the beginning of *The Order of Things*, I am interested in reflecting on the fact that Latin America is, for Huntington, a civilization in itself and not part of the Western Hemisphere. For Huntington, Latin America has an identity that differentiates it from the West:

> Although the offspring of European civilization, Latin America has evolved down a very different path from Europe and North America. It has a corporatist, authoritarian culture, which Europe has to a much lesser degree and North America not at all.[39]

Apparently, Huntington does not see fascism and Nazism as authoritarian. Nor does he perceive the fact that U.S. authoritarianism, since 1945, has projected control of international relations through a new form of colonialism: colonialism without territoriality. However, Huntington invokes even more traits to mark Latin American difference:

> Europe and North America both felt the effects of the Reformation and have combined Catholic and Protestant cultures. Historically, although this may be changing, Latin America has been only Catholic.[40]

At this point in the argument, the difference invoked is the imperial difference that the Reformation initiated and subsequently took form, beginning in the eighteenth century with the development of science and philosophy, and especially in the concept of "Reason" that brought coherence to the discourse of the second modernity. Moreover, the third important component of Latin America is, in Huntington's view, "the indigenous cultures, which did not exist in Europe, were effectively wiped out in North America, and which vary in importance from Mexico, Central America, Peru and Bolivia, on the one hand, to Argentina and Chile on the other."[41] Here, Huntington's argument passes from the imperial difference to the colonial difference, as much in its originary form in the sixteenth and eighteenth centuries as during the nation-building period, which is precisely where the difference between Bolivia and Argentina, for example, becomes evident—when the national model is imposed from Northern Europe on to the former

Coloniality at Large • 67

Hispanic empire. In conclusion to these observations, Huntington maintains:

> Latin America could be considered either a subcivilization within Western civilization or a separate civilization closely affiliated with the West. For an analysis focused on the international political implications of civilizations, including the relations between Latin America, on the one hand, and North America and Europe, on the other, the latter is the more appropriate and useful designation.... The West, then, includes Europe, North America, plus the other European settler countries such as Australia and New Zealand.[42]

About what is Huntington thinking when he speaks of "other European settler countries such as Australia and New Zealand"? Obviously, he is thinking about English colonialism in the second modernity, in the imperial difference (the English colonialism that "surpassed" Iberian colonialism) mounted over the colonial difference (certain colonial heritages belong to the West; others do not). In the colonial heritages that belong to the West, the indigenous component is ignored, and, for Huntington, the strength that indigenous movements are acquiring in New Zealand and Australia does not appear to be a problem. Nevertheless, the panorama is clear: The West is the new designation, after the Cold War, for the "first world." The "West" has become the locus of enunciation that produced and produces imperial and colonial differences, the two axes around which the production and reproduction of the modern/colonial world turn. Although the emergence of the idea of the "Western Hemisphere" offered the promise of an inscription of the colonial difference from colonial difference itself, the "Roosevelt corollary" instead reestablished the colonial difference from the north and through the definitive defeat of Spain in the Spanish-American War. The fact is that Latin America today, in the new world order, is a product of the originary colonial difference and its rearticulation over the imperial difference that gestated from the seventeenth century in Northern Europe and was constituted in the emergence of a neocolonial country like the United States.

What importance can these geopolitical abstractions have in the reorganization of the global order in a hierarchical order of civilizations like the one Huntington proposes? Let us point out at least two: on the one hand, international relations and the economic order of the future; on the other hand, the migratory movements and public politics of the countries see themselves as "invaded" by habitants of "non-Western civilizations." In the first case, the question is that to maintain, in Huntington's terms, a unity like Latin America means conferring to it a place in international alliances and the concentration of economic power. In the second place, it directly affects growing Latin American immigration toward the United States, which has some 46 million "Hispanics" as of 2007. Let us look at these two aspects in more detail, although in a somewhat brief form.

The end of the Cold War and the fall of the socialist world brought about new theories that predicted a future world order, as much as in the economic realm as in the arena of civilization. Huntington's need to establish a world order based on civilizations answered his fundamental thesis that the wars of the future would be wars between civilizations more than ideological wars (such as the Cold War) or economic wars (such as the Gulf War). Immanuel Wallerstein predicted that the new world order would coalesce between 1990 and 2025/2050.[43] In Wallerstein's scenario, there are several reasons for a coalition between the United States and Japan. In such a situation, the European Union would be a second powerful group, yet different from the first. In this scenario, two countries of enormous human and natural resources remain in an uncertain position: Russia and China. Wallerstein predicts that China would come to form part of the U.S.– Japanese coalition, while Russia would ally itself with the European Union. The possibility that this scenario could come to pass offers interesting possibilities to reflect on the rearticulation of the imaginary of the modern/ colonial world, that is to say, the rearticulation of the coloniality of power and the new global colonialism. The possible alliance between the United States, on one side, and China and Japan, on the other, would mean a 180-degree turn over the last 600 years: The emergence of the Atlantic commercial circuit was, in the sixteenth century, one of the consequences of the strong attraction that China offered (as a function of the commercial marginality of Europe). At the end of the economic, cultural, and ideological consolidation of the Atlantic, there would be a re-meeting of the colonial difference in one of its geohistorical locations (e.g., the Jesuits in China).[44] The reorganization and expansion of global capitalism would produce a meeting between Chinese civilization (in Huntington's broad meaning, from 1500 B.C. until the current communities and countries of the Asian southeast, such as Korea and Vietnam) and Western civilization, or at least part of it.[45] In reality, one of the interests of Wallerstein's scenario was to suppose that Western civilization would be divided: Part of it would establish alliances with the Chinese and Japanese civilizations (or two aspects of the same civilization) and the other (the European Union) with one of the margins of the West, or what Huntington calls "the Russian orthodox civilization," which differs from its close relatives, the Byzantine and Western civilizations.⁴⁶ A fascinating scenario, in truth, since the imaginary of the modern/colonial world that accompanied and justified the history of capitalism was a point of radical transformations. That is to say, capitalism would enter a phase in which the initial imaginary would disintegrate into other imaginaries, or better, that capitalism *is* the imaginary and, consequently, that Huntington's different civilizations would be destined to be pulverized by the intransigent march of the exploitation of labor at the national and transnational levels.

Six years after Wallerstein's predictions, the magazine *Business Week* (February 8, 1999) asked, in a boldfaced headline, "Will it be the Atlantic Century?" In smaller red letters, in the same headline, they suggested an answer: "The 21st Century was supposed to belong to Asia. Now the U.S. and Europe are steadily converging to form a new Atlantic economy, with vast impact on global growth and business."⁴⁷ This scenario should come as no surprise. The colonial difference is redefined in the global forms of colonialism motivated by finance and the market more than by Christianization, the civilizing mission, Manifest Destiny, or progress and development. What is surprising is Wallerstein's scenario. However, the only problem that attracts attention is the question, "Will it be the Atlantic century?" referring to the twenty-first century, of course. The question attracts attention for the following reason: Was it not the case that the last five centuries have been the Atlantic centuries? But the emphasis here is not on the Atlantic, but on the North Atlantic, the new geopolitical designation in an imaginary that replaces the differences between Europe and the Western Hemisphere with the emergence of the North Atlantic. Certainly, this scenario did not escape Huntington when, while redefining the West, he affirmed, "Historically, Western Civilization is European civilization. In the modern era, Western civilization is Euroamerican or North Atlantic civilization. Europe, America [and I would say North America] and the North Atlantic can be found on a map, the West cannot."⁴⁸ With the disappearance of the West, the Western Hemisphere also disappears. As Kissinger foresaw in the paragraph cited at the beginning of this article, the Western Hemisphere only remains as a question "internal" to North America in the rearticulation of the colonial difference in the period of global colonialism.

> The second consequence mentioned earlier is the status of the south-to-north migrations that are producing the "Latin Americanization" of the United States. If the "Roosevelt corollary" was a triumph of the consciousness and power of the Anglo-American over the consciousness and power of the Latin American, the massive migrations from

south to north are evincing a new dimension, which is reinforced in social movements. The migrations not only include white Latinos and mestizos, but also numerous indigenous persons who have more in common with Native Americans in the United States than with whites or mestizos in Latin America.[49] On the one hand, due to the politics of the United States in the Caribbean in its moment of expansion before World War II, Afro-American immigration from Jamaica and Haiti complicates the scenario at the same time as it throws into relief the silenced dimension of the white Creole- and mestizo-controlled north-south relations established with the idea of the Western Hemisphere. For indigenous and Afro-American populations, the image of the Western Hemisphere was not—and is not—significant. This is one of the aspects to which Huntington refers when he says: "Subjectively, Latin Americans themselves are divided in their self-identification. Some say, "Yes, we are part of the West." Others claim, "No, we have our own unique culture."[50]

Both positions can be sustained from the perspective of Creole double consciousness in Latin America. It would be more difficult to find evidence that these opinions have their origin in indigenous or Afro-American double consciousness. And this distinction is not only valid for Latin America, but for the United States as well. Huntington attributes to Latin America a "reality" that is valid for the United States, but perhaps is not perceptible from Harvard, since from there, and from the connections political and social scientists have with Washington D.C., the gaze is directed more eastward (London, Berlin, Paris) than toward the Southeast and Pacific, which are residual spaces, spaces of the colonial difference. However, while at Harvard, W. E. B. Du Bois could see the South and understand that for those who are historically and emotionally linked to the history of slavery, the question of being Western or not is not put forth.[51]

And if the problem is introduced, as in the recent book by the Caribbean-British intellectual Paul Gilroy, it appears in an argument in which the "Black Atlantic" emerges as the forgotten memory, buried under Huntington's "North Atlantic."[52] On the other hand, the reading of the eminent Native American intellectual and lawyer from the Osage community, Vine Deloria Jr., demonstrates that indigenous communities in the United States were not totally eliminated, as Huntington asserts.[53] Furthermore, Deloria shows that the colonial difference that emerged with the imaginary of the Atlantic commercial circuit persists in the United States, and that it was necessary for the historical foundation of Western civilization and its internal fracture with the emergence of the Western Hemisphere. There is much more to Deloria's argument than the simple difference between Protestantism and Catholicism that preoccupied Huntington. Deloria reminds those who have a bad memory of the persistence of the forms of knowledge that not only offer alternative religions, but more important still, alternatives to the concept of religion that is fundamental to the architecture of the imaginary of Western civilization. The transformation of the "Western Hemisphere" into the "North Atlantic" secures, on the one hand, the persistence of Western civilization. On the other hand, it definitively marginalizes Latin America from Western civilization, and creates the conditions for the emergence of forces that remain hidden in the Creole (Latin and Anglo) of the "Western Hemisphere"—that is, the rearticulation of Amerindian and Afro-American forces fed by the growing migrations and technoglobalism. The Zapatista uprising, the force of the indigenous imaginary, and the dissemination of the Zapatista discourses have made us think about possible futures beyond the Western Hemisphere and North Atlantic. At the same time, beyond all civilizatory fundamentalism (ideological or religious), whose current forms are the historical product of the "interior exteriority" to which they were relegated (e.g., subalternized) by the self-definition of Western civilization and the Western Hemisphere, the problem of the "Westernization" of the planet is that the whole planet, without exception and in the last 500 years, has had to respond in some way to Western expansion. Therefore, "beyond the Western Hemisphere and North Atlantic," I don't want to say that there exists some "ideal place" that must be defended, but merely that there necessarily is something "beyond" global organization based on the interior exteriority implied in the imaginary of Western civilization, the Western Hemisphere, and the North Atlantic.

Notes

A preliminary version of this paper was delivered at the meeting of the International Congress of Sociology in Montreal, August 1998, in a special panel, "Alternatives to Eurocentrism and Colonialism in Latin American Social Thought." The panel was organized by Edgardo Lander and Francisco Lopez Segrera under the auspices of the UNESCO. A second version was delivered as a lecture in the Department of History, Princeton University, spring 1999.

1. Henry Kissinger, *Years of Renewal* (New York: Simon and Schuster, 1999), 703.
2. Aníbal Quijano and Immanuel Wallerstein, "Americanity as a Concept, or the Americas in the Modern World-System," *International Social Sciences Journal* 134 (1992).

3. Edouard Glissant, *Poetics of Relation*, trans. Betsy Wing (Ann Arbor: University of Michigan Press, 1997).
4. Immanuel Wallerstein, *The Modern World-System: Capitalist Agriculture and the Origins of the European World-Economy in the Sixteenth Century* (New York: Academic Press, 1974).
5. Frank A. Gunder and Barry K. Gills, eds., *The World System: Five Hundred Years or Five Thousand?* (London: Routledge, 1993).
6. Giovanni Arrighi, *The Long Twentieth Century* (London: Verso, 1994).
7. Aníbal Quijano, "Colonialidad del poder, cultura y conocimiento en América Latina," *Anuario Mariateguinao* 9 (1997): 113–121; Walter D. Mignolo, "Colonialidad del poder y diferencia colonial," *Anuario Mariateguiano* 10 (1999); Walter D. Mignolo, *Local Histories/Global Designs: Coloniality, Subaltern Knowledges and Border Thinking* (Princeton: Princeton University Press, 2000).
8. Immanuel Wallerstein, *Geopolitics and Geoculture: Essays on the Changing World-System* (Cambridge: Cambridge University Press, 1991); Wallerstein, "The French Revolution as a World-Historical Event" in *Unthinking the Social Sciences: The Limits of Nineteenth-Century Paradigms* (Cambridge: Polity Press, 1991); Wallerstein, "The Geoculture of Development, or the Transformation of Our Geoculture" in *After Liberalism* (New York: New Press, 1995).
9. Quijano and Wallerstein, "Americanity as a Concept, or the Americas in the Modern World-System;" Quijano, "Colonialidad del poder, cultura y conocimiento en América Latina;" Quijano, "The Colonial Nature of Power and Latin America's Cultural Experience," in Roberto Briceño León and Heinz R. Sonntag, eds., *Social Knowledge: Heritage, Challenges, Perspectives*, Vol. 5, International Sociological Association, Pre-Congress Volumes (1998).
10. Arrighi, *The Long Twentieth Century*.
11. Quijano and Wallerstein, "Americanity as a Concept, or the Americas in the Modern World-System," 449.
12. Janet L. Abu-Lughod, *Before European Hegemony: The World System a. d. 1250–1350* (New York: Oxford University Press, 1989); Eric C. Wolf, *Europe and the People Without History* (Berkeley: University of California Press, 1982); Mignolo, *Local Histories/Global Designs*.
13. Bernard Lewis, *The Shaping of the Modern Middle East* (New York: Oxford University Press, 1997.
14. Rolena Adorno, *Writing and Resistance in Colonial Peru* (Austin: University of Texas Press, 1986); Serge Gruzinski, *La colonization de l'imaginaire: Sociétés indigenes et occidentalisisation dans le Mexique espagnol XVI–XVIII siècle* (Paris: Gallimard, 1988); Enrique Florescano, *Memory, Myth, and Time in Mexico* (Austin: Austin University Press, 1994); Sabine MacCormack, *Religion in the Andes: Vision and Imagination in Early Colonial Peru* (Princeton: Princeton University Press, 1991).

15. Leslie Marmon Silko, *Almanac of the Dead* (New York: Simon and Schuster, 1991).
16. Rigoberta Menchú, "Los pueblos indios en América Latina," in Pablo González Casanova, *Democracia y Estado multiétnico en América Latina* (Mexico: UNAM, Centro de Investigaciones Interdisciplinarias, 1996), 125.
17. W. E. B. Du Bois, *The Souls of Black Folk* (New York: Vintage Books, 1990).
18. Rigoberta Menchú, *Me llamo Rigoberta Menchú y así me nació la conciencia* (London: Verso, 1982); Gloria Anzaldúa, *Borderland/La Frontera: La Nueva Mestiza* (San Francisco: Aunt/Lutte, 1987).
19. Du Bois, *The Souls of Black Folk*, 8–9.
20. Laura Bornholdt, "The Abbi de Pradt and the Monroe Doctrine," *The Hispanic American Historical Review* 24 (1944): 201–221.
21. Thomas Jefferson, *The Writings of Thomas Jefferson*, Vol. 13, ed. A. A. Lipscomb (Washington D.C.: Library of Congress, 1903–1904), 22.
22. Simón Bolívar, "Carta de Jamaica," in comp. Leopoldo Zea, *Fuentes de la Cultura Latinoamericana*, Vol. 1 (Mexico: Fonda de Cultura Económica, 1993), 25.
23. Bolívar, "Carta de Jamaica," 17.
24. Jefferson, *The Writings of Thomas Jefferson*, 24.
25. Michel-Rolph Trouillot, *Silencing the Past: Power and the Production of History* (Boston: Beacon Press, 1995).
26. Dana Nelson, *National Manhood, Capitalist Citizenship and the Imagined Fraternity of White Men* (Durham: Duke University Press, 1998).
27. Jefferson, *The Writings of Thomas Jefferson*, 12.
28. Joan Dayan, *Haiti, History and the Gods* (Berkeley: University of California Press, 1998), 19–25.
29. Jorge Klor de Alva, "The Postcolonial of (Latin) American Experience: A Reconsideration of 'Colonialism,' 'Postcolonialism,' and 'Mestizaje,'" in Gyan Prakash, *After Colonialism: Imperial Histories and Postcolonial Displacements* (Princeton: Princeton University Press, 1995).
30. Arthur P. Whitaker, *The Western Hemisphere Idea: Its Rise and Decline* (Ithaca: Cornell University Press, 1954), 87–100.
31. Whitaker, *The Western Hemisphere Idea*, 100.
32. Whitaker, *The Western Hemisphere Idea*, 100.
33. Whitaker, *The Western Hemisphere Idea*, 107.
34. Wallerstein, *Geopolitics and Geoculture*.
35. Wallerstein, *Geopolitics and Geoculture*, Wallerstein, "The French Revolution as a World-Historical Event"; Wallerstein, "The Geoculture of Development, or the Transformation of Our Geoculture."
36. Trouillot, *Silencing the Past*.
37. Quijano, "Colonialidad del poder, cultura y conocimiento en América Latina"; Mignolo, *Local Histories/Global Designs*.

74 • Walter D. Mignolo

38. Wallerstein, "The Geoculture of Development, or the Transformation of Our Geoculture," 1166.
39. Samuel Huntington, *The Clash of Civilizations and the Making of World Order* (New York: Simon and Schuster, 1999), 46.
40. Huntington, *The Clash of Civilizations and the Making of World Order*, 46.
41. Huntington, *The Clash of Civilizations and the Making of World Order*, 46.
42. Huntington, *The Clash of Civilizations and the Making of World Order*, 47.
43. Wallerstein, "The Geoculture of Development, or the Transformation of Our Geoculture," 32–35.
44. Jonathan Spence, *The Chan's Great Continent: China in Western Minds* (New York: W. W. Norton and Co., 1999).
45. Huntington, *The Clash of Civilizations and the Making of World Order*, 15.
46. Huntington, *The Clash of Civilizations and the Making of World Order*, 45.
47. *Business Week*, cover image February 8, 1999.
48. Huntington, *The Clash of Civilizations and the Making of World Order*, 47.
49. Stefano Varese, *Pueblos indios, soberanía y globalismo* (Quito: Biblioteca Abya-Yala, 1996).
50. Huntington, *The Clash of Civilizations and the Making of World Order*, 47.
51. Du Bois, *The Souls of Black Folk*.
52. Paul Gilroy, *The Black Atlantic: Modernity and Double Consciousness* (Cambridge, MA: Harvard University Press, 1993).
53. Vine Deloria, *God Is Red: A Native View of Religion* (Colorado: Fulcrum Publishing, 1993).

PART 2

Disciplining Hemispheric Studies

CHAPTER 4

A Major Motion Picture: Studying and Teaching the Americas

Michael O. Emerson

My spouse grew up in Minnesota. I moved there when I was five, as my parents wanted to move back to the state they knew best. Although my spouse, children, and I have lived in Texas for over a decade now, nearly all the rest of our families live in Minnesota. At least once a year, and usually two to three times, we drive to Minnesota (about 1,300 miles) to visit family, attend family functions, and check on old friends. Our children were all raised in Texas, but because of their regular trips to Minnesota, because Minnesota is where their relatives live, and because Minnesota even has sports teams for which they cheer, they consider Minnesota their second home. When they talk about where they want to live when they grow up, they usually mention Texas and Minnesota in the same breath, as if the states geographically border each other and as if the vast differences between the states did not matter.

In a small way, we contribute to a pipeline between Texas and Minnesota, with members of my immediate family traveling up and down it at least yearly. Each trip—no matter the direction—brings with it a bit of culture from the other location, usually materials. Minnesota relatives have all at one time or another made the trek down to Texas to visit. Children bring Texas friends with them to Minnesota to show them snow and cold. Two families of our Minnesota relatives decided to move to Texas, and one Texas family that visited has since decided to move to Minnesota.

Though I cannot know for sure, my bet is that most readers of this essay can tell a similar story, substituting their own states or nations. And this leads me to the point of this essay. We can go far in understanding, researching, and teaching the Americas with a simple conceptual tool: *motion*. This essay will outline the concept of motion, examine one type of motion in more detail—body motion, that is, migration—and explore how we can use this concept as one unifying approach to further study the Americas, to teach our students, and to help them expand their learning. The payoff in using "motion," rather than "migration" or other terms is that it is meant to capture movements of all sorts, not simply the migration of people. By using this particular term, we better capture what is the Americas.

For 500 years, the Americas have been defined by their motion. People moving to the Americas, moving back and forth across borders, cultures being shared, resources and products moving along well-defined trade routes (often in exploitative fashion), political ideas and educational systems being shared, formally separate groups of people intermixing, and religious ideas and practices moving in a dizzying array of directions across the Americas, such as in this example (Levitt 2004:1):

> Every Sunday morning, a group of families in Governador Valadares, Brazil, gather in their living room to watch the Catholic Mass that is broadcast on their local TV. But this mass is not held in Valadares or any other Brazilian city. Instead, it is a videotaped recording of the Portuguese mass held at St. Joseph's Church in Somerville, Massachusetts where large numbers of Brazilians have migrated. Family members still living in Brazil watch hoping they will glimpse their relatives worshipping in Boston. As they watch for their relatives and follow the liturgy, change comes to Governador Valadares. American priests come to be known and respected, bits of English are learned, and subtly different ways of celebrating mass are communicated. With technology, motion no longer has to be carried literally by people, as in my opening Texas/Minnesota example. The television, the Internet, and a host of other technologies not only make communication easier, they also make motion itself easier. Ideas, products, politics, religion, cultural sensibilities, and so much more traverse between peoples instantly.

I use the concept of motion rather than the oft-used term transnationalism because I mean something beyond transnationalism. Transnationalism means the flow of people, information, goods, services, and other resources across national boundaries. By definition, flows within boundaries are not part of transnationalism. The term also is rather sterile, having come to mean macro-level, large-scale flows, with larger-sized flows indicating greater transnationalism.

By using the concept of motion, I mean to (a) create a more descriptive term of the actual processes at work, (b) not presuppose that only movement across national borders is what matters (borders change after all, and secondary movements within nations also matter), and (c) imply that we are not just talking about large-scale flows across nation-states, but also of ways of being that characterize the Americas. Thus, in using motion as an organizing concept by which to understand the Americas, I wish to focus attention directly on the broad array of movement that is the Americas—the motion of people (moving across borders, yes, but also changing individually and culturally, moving within new borders to reconnect or for other reasons), of capital, of cultures, of ideas, of technology, of history.

Motion and movement across the globe is an increasingly defining characteristic of modern times. But the level of motion and degree of shared collective memory and contemporary experience suggest that motion is a particularly useful concept for understanding the Americas. As we have seen in earlier chapters (for example, see Chapter 2 by Caroline Levander), the Americas share a common heritage; are spatially separated from other continents by vast oceans (which traditionally meant less motion across those oceans); are not nationalistic in the way many nations are because they are internally diverse and pluralistic, relatively young, and have experienced dramatically fewer separatist movements; and ultimately are united in the collective ideal known as the "New World."[1]

The new world shares a European heritage. The dominant languages of the Americas—Spanish, English, Portuguese, and French—are all European languages, reflecting the common past of European colonialism. But at the same time, the Americas are a mixture of native peoples, European descendents, and African descendents. To varying degrees, the Americas share in the European "disease" of racial and ethnic stratification, and it is a disease that by most accounts reaches its zenith in the new world.[2] And the Americas are even linked in their hierarchy and systems of exploitation. Motion, of course, does not imply simply neutral or positive interactions or movement. The Americas are a unique mixing of peoples, with such terms as mestizaje, multiracial families, multiethnic identity, Creole, and rainbow people used to describe the motion of groups of people intermixing. Ultimately, "what is happening in the Americas does not easily conform to globalization or Europeanization.... [they are] dissimilar cultural dynamics."[3]

Before we go further into our analysis of the Americas and motion, let us consider the basic structure of the Americas. It is two continents—North and South America—and includes island-nations off the coast of the main continents (for example, Cuba, the Bahamas, and Jamaica). Collectively, about one billion people live in the Americas—a lot of people, to be sure, but not as many people as live in India, and not as many people as live in China.

The Americas are divided into approximately forty-five nation-states. One of those nation-states—the United States—contains about one-third of the Americas' population, and has for at least 100 years been the dominant military, economic, political, and cultural force in the region, often to the consternation of others. Two other nations—Brazil and Mexico—contain another third of the Americas' population. Each of these three nations speaks a different language, and can be thought of as the cultural centers of the English-, Portuguese-, and Spanish-speaking portions of the Americas, respectively. The remaining one-third of the Americas' population is scattered across the more than forty other nation-states in the hemisphere.[4]

Examining the social structure of the Americas is part of a sociological approach. A sociological approach to the study of the Americas can bring several benefits to the conversation. Sociology considers the ways in which individuals, their groups and societies, and their histories interact to shape demographics, stratification, cultures, and further social relations of all types. It also brings with it a set of methodologies—qualitative and quantitative—and theory that guides such study. This essay demonstrates some of the potential in such analysis.

Inequality, Development, and Body Motion

At the core of motion in the Americas, at least traditionally yet still of vital importance today, is the physical moving of humans across and within national boundaries. Why is there migration from one place to another? Ask the proverbial "average persons" on the street and they will probably say it occurs because people are trying to make a better life. They are partly correct. Such an answer, though, begs the question—why is life better somewhere else, and why move now (as opposed to before or later)?

As Doug Massey (1999) identified, immigration streams begin largely out of economic circumstances—all else being equal, the greater the economic inequality between two places, the more people will move from the economically challenged place to the place of economic opportunities, but as we will see, usually only in the case of displacement and economic and social upheaval. Thus, the initial wave of immigration typically is economically driven. Once such a migration path starts, it greases the wheel for more migration along the same route. In fact, research finds that migration streams overshoot economic opportunity, and this is because of the second wave of immigration. That is, even when jobs are no longer plentiful or economic benefits clear-cut, migration continues due to social reasons.[5] People continue to move to reunite with family, or because they have come to feel like the new land is their second home, or they view the new land as a place they ought to go to because so many others have, or even because they seek a new culture and new experiences.

People move for other reasons, too, of course, though not always of their own free will. Persecution, political instability, famine, natural disaster, or other calamities may push them to emigrate, though again, the intent is still to make a better life (or, in some cases, simply to preserve life). Thus, migration streams ultimately come and go. When there was a potato famine in Ireland, many people emigrated. And following the "two wave logic" mentioned earlier, the Irish continued to emigrate well after the potato famine. But eventually, with no strong push factors and with no dramatic economic disparities between Ireland and the wealthiest nations, emigration slowed measurably. The physical movement of people along set paths is not continuous, but shaped in significant ways by the comparative fortunes and power of nations and regions. Yet the dynamic transmission of ideas, modes of being, technologies, and cultures can continue indefinitely, long after physical motion, or even apart from such movement.

And this leads us to another often-overlooked fact. What is defined as international migration or even transnationalism also is dependent upon political boundaries. I have an acquaintance whose family has lived for the past ten generations in Texas on the Rio Grande. Across these generations, his family has been under the flag of France, Spain, Mexico, the Republic of Texas, and the United States. Yet many in his family have never moved off the ranch in all these generations. They did not move; the boundaries did. So we see motion also in political boundaries, contested through wars and treaties. The physical bodies can be static, yet there is motion.

A final type of immigration is enforced migration—the capturing and enslaving of peoples. The Americas, more than anywhere else in the world, have millions of descendants of forced migrants. Let us look at the periods of "body motion" to and within the Americas to develop these ideas more fully.

Migration Eras

During the Mercantile Period (1500–1800), world immigration flows were dominated by Europe. Specifically, as Europeans colonized the world some of them moved to the Americas. This period also was the era of slave trade to the Americas, which helped make the Americas economically productive, and, where possible, enslavement of native peoples, which helped to produce a common heritage (ugly as it is) that the Americas largely share.

The Industrial Period (1800–1925) found immigration to the Americas still shaped by what was occurring in Europe (and to a much smaller degree in Asia). The type of economic development in Europe during this period—the Industrial Revolution—displaced people from their agrarian lives. In search of ways to survive and thrive, more than 48 million people left Europe for the Americas and Oceania. At least three-quarters went to Argentina, Canada, and the United States, with the United States alone receiving 60 percent of the European immigrants during this period.

Of course, the language of immigrants often determined which nations in the Americas they went to. Europeans speaking Spanish or Italian moved to South and Central America. Many Portuguese-speaking immigrants moved to Brazil. French speakers moved to Quebec in Canada and several islands of the Americas. English speakers largely ended up in the United States and Canada, and on some of the islands of the Americas. What is important to note is that the movement of people to the Americas during this period was overwhelmingly European. For example, during this period, nearly 90 percent of all immigrants to the United States were European (forced African immigration ended early in this period). Immigration between nations of the Americas occurred, but was relatively small in proportion to the inflow of Europeans.

The world wars and the Great Depression essentially stopped immigration for several decades, but a new period of people motion began in full force in the 1960s, continuing to this day. Some call this the postindustrial period of migration. Instead of movement from Europe to former colonies, immigration has globalized and is dominated by migration from poor, but economically developing, nations to wealthy nations, which, almost without exception, have below replacement birth rates, thus necessitating the importation of workers. It is not poverty itself—either at the individual/family or national level—that leads to migrations to wealthier areas. Ironically, it is only when economic development occurs. As migration scholar Douglas Massey (1999:48) writes:

> International migration originates in the social, economic, cultural, and political transformations that accompany the penetration of capitalist markets into nonmarket and premarket societies.... [such a pattern] disrupts existing social and economic arrangements and brings about a displacement of people from customary livelihoods, creating a mobile population of workers who actively seek for new ways of achieving economic sustenance.

Migration then does not come from lack of economic development. Rather, it comes from development itself.

The recent decades have been a period of substantial economic development across the Americas, especially in Latin American nations. Poverty rates have been reduced, per capita income has increased, food production is up. Yet it is these very processes (and their unevenness in distributing the benefits across regions and people) that have led to dramatic increases in the movements of people across the Americas, both to urban areas within their own nations and across national borders. Migration in this sense is not undertaken to make a better life per se, but "as a means of overcoming market failures that threaten their material well-being..." (Durand and Massey 2004:48). Ultimately, to fully understand the motion of people across space one must consider (a) the treatment of large, structural forces that push and pull people; (b) the motivations, goals, and aspirations of people; (c) the economic and social forces that arise to connect the areas of in and out migration; and (d) the nation-state policies to control the flow of people.

Consistent with the global pattern of immigration in the modern period, this body motion within the Americas has overwhelmingly been from developing regions to already wealthy nations and regions within the Americas—the United States, Canada, parts of Brazil and Argentina—and to the most developed cities of many of the nations of the Americas (with much smaller flows to the former colonizer nations in Europe). For example, since the 1960s, rural people have moved to urban areas of Argentina and Venezuela, not only from those nations, but also from Bolivia, Brazil, Paraguay, and Uruguay.[6]

Also consider the immigration of Mexican peoples to the United States. Since about the 1970s, their nation has been in the throes of expansive economic development and market integration. In 1950, the Mexican-born population in the United States was 500,000. By 1960, the number—600,000—had barely changed from ten years earlier. A bit more growth occurred in the number of Mexican-born persons in the United States by 1970: 800,000. But the 1970s saw Mexico enter full force into the global market economy, and it is reflected in the growth of the number of Mexican-born persons in the United States since then: In 1980, 2.2 million persons; in 1990, 4.5 million persons; in 2000, 9.2 million persons; and in 2010, 14 million persons.[7] The number of Mexican-born persons living in the United States is greater than those who live in all but the eight most populous nations of the Americas. [Interestingly, the movement of people flows in the opposite direction as well, though on a much smaller scale. According to the 2000 Mexican census nearly 350,000 Americans have legally immigrated to Mexico, and if undocumented persons are included, some argue that as many as 1 million Americans live in Mexico.[8]] And this trope of motion functions as a way of framing other developments in the Americas—the movement of ideas, cultural forms (e.g., music), etc. Many studies have examined such developments; many more studies are needed before we have anything approaching full understanding of the myriad connections between forms of motion.

For approximately twenty years, the Mexican Migration Project has followed the migration of Mexicans surveyed in eighty-one Mexican communities and U.S. branch settlements. That is, Mexican communities have been surveyed for twenty years, and the migrants are also surveyed in their U.S. destinations. The findings are important for understanding the motion of people.

The project discovered that much of the movement of Mexicans to the United States is "transitory, circular, and clandestine and thus outside the purview of normal statistical systems."[9] A full 60 percent of the migrants in their sample were undocumented in their most recent move to the United States. The researchers also found that it is not the rural poor looking to survive who migrate to the United States. Rather, they conclude that probably the most important motivation for the migration is the need to self-finance home acquisition because of inaccessible and poorly operating mortgage markets in Mexico. The key motivation of many migrants, then, is not to engage in one-way motion; it is to move resources from the United States back to Mexico, thus necessitating moving one's self, temporarily, to the United States and then back. The motion of people is often circular. Return migration, then, is common; between 40 and 55 percent of Mexican immigrants to the United States return to live in Mexico within three years.[10] In terms of home acquisition and improvement in Mexico, this strategy apparently works for many:

> In general, the greater a household's prior U.S. experience, the higher the odds it will own its home, the greater the number of rooms and appliances in the dwelling, and the more likely the dwelling will be to have a tile or wood floor.[11]

The circular body motion to the United States and back leads to further types of motion, the researchers find. Gender relations change, as the men go to the United States or the larger cities of Mexico, and the women left behind take on more central roles in the household. U.S. culture, ways of thinking and doing, and consumer goods find their way back to their Mexican communities, even as the United States is transformed by the many millions of Mexican and other immigrants (salsa outsells ketchup in the United States, for instance). In what is called the hypothesis of cumulative causation, the changes that come from migration lead to even more migration: "the departure of people and the repatriation of earnings change local social and economic structures in ways that promote additional migration."[12] Motion begets motion begets motion.

The physical movement of peoples, then, is associated both with more physical movements of people *and* with a host of other movements—religion, knowledge bases, cultural practices and products, collective memories, world views, language, and social ties. To study and understand these associated movements is not only important, but also challenging and profoundly rich. Much work in hemispheric studies has focused on aspects of these movements, but so much more remains to be done. Most notably, we need to better connect these associated movements and to theorize how they are related to one another. As scholars Sandya Shukla and Heidi Tinsman write in the opening of their book, *Imagining the Americas*:

> [we ought to be] interested not so much in "comparative history"—the side-by-side examination of different "countries"—but in the experiences, imaginaries, and histories of interaction. These spaces of dialogue, linkage, conflict, domination, and resistance take shape across, or sometimes outside, the confines of national or regional borders and sensibilities and therefore allow for new epistemologies. Shared problematics, then, rather than a common geography, colonizing power, or language, might define an "American" inquiry...[13]

Inhibitors to Motion

But motion is not free-flowing and devoid of resistance. We make light of the struggle and turmoil that constantly ensues with such motion if we ignore these realities. Let us consider the reality of borders, which though politically and socially constructed, play significant roles in the identities of the Americas and in the motion of people, resources, and more. Borders are not arbitrary. They are the result of real political struggles and negotiations, of victories and defeats. They are contested, of course, and can change. To hold them requires attention—military, rhetoric, symbolic. These processes reify the importance of borders and amplify distinctions between the nation-states of the Americas and their peoples. The borders become the stage by which people of all sorts—journalists, ideologues of every type, politicians, special interest groups—play out their fears, hopes, and identities about nations. Immigrants—those in body motion—become mere symbols in a larger battle of definition, identity, and power struggles.

The results vary, of course, but for the United States, such battles and debates have led to a series of what can be seen as repressive immigration and border policies—for example, fences being built along the border, the expansive increase in Border Patrol officers and technology in an attempt to limit the motion of people across nation-state borders, and in 1986, the passage of the Immigration Reform and Control Act, which criminalized the hiring of undocumented workers.

Studies find that the result of these body motion inhibitors, rather than deterring migration, significantly decreased the likelihood of return migration, and thus the number of undocumented people in the United States has significantly increased. Given the greater costs to migrating across the border, the proportion hiring a coyote has increased. What is more, human-trafficking markets have taken fuller root, given the continued need for workers from Mexico but within the context of more restrictive immigration policies.

Just a few days before writing this section, I met with some representatives of the local-area Catholic Charities headquarters in Houston, Texas. They told me that within the city they were hiding hundreds of people from the traffickers who "owned the immigrants." And these immigrants were not just adults. Nearly every week, they said, they receive youth and children who have been trafficked across borders to work as servants in the United States. The number of people coming to their offices has increased exponentially since the changes in immigration policies and the increased border patrols. Motion inhibitors may not always succeed in reducing motion, but they do change its nature and increase its costs. Lives are affected in profound ways in these battles over borders and national identities. While motion defines the Americas, much time is spent pretending it just isn't so. Partly as a result, people die in the process of attempting these moves. Much more remains to be studied when considering motion inhibitors. Laws, policies, the processes by which they are created and enacted, and how they affect relations between nations, communities, and people all need substantially closer examination. Study of the diaries, letters, art, and other communication of migrants as they face these inhibitors have begun, but we have only scratched the surface of what is available. Rich literatures have arisen historically in response to such inhibitors, but again, much work awaits us if we are fully to understand their meanings.

Filming a Motion Picture

With the concept of motion in hand, and with a closer look at the perhaps most visible manifestation of motion—that of people moving to different regions and back—let us now develop an extended example. How can we study the motion of the Americas without becoming overwhelmed by its vastness? How can we use the concept of motion in concrete ways to teach our students about the Americas?

Let's focus on one neighborhood in Houston. In fact, this neighborhood was where my family and I lived from 1999 until we moved to Chicago for a time in 2004 (we have since returned to Texas). This neighborhood was a new subdivision on the outer part of Houston, and as such, there were no clear understandings of whose neighborhood it was or would be. This, in part, explains the ethnic, national, and racial diversity of the neighborhood. Also, the homes were affordable to most working- and middle-class people, costing from $75,000 to $150,000.

Now I will admit that not many people would actually take the time to make a chart of the homes in their neighborhood and record the ethnicities of everyone living in those homes. But I am admittedly odd, and did just that. I present the results for my immediate block in Figure 4.1.

This cul-de-sac type block of twenty-two homes has a wide variety of people, about three-quarters of whom are from the Americas, including El Salvador, Guatemala, Haiti, Jamaica, Mexico, Nicaragua, Puerto Rico, and the United States. Immigrants from other parts of the world—China, India, Korea, and Nigeria—also live on the block. As the figure shows, my family and I live in the home with the arrow, between the Asian Indian family and the Guatemalan family.

As scholars, we and our students could find enough research projects on this one block to fill a lifetime of study about the Americas and its

Figure 4.1 Neighborhood in 1999

motions. An initial question my students ask when I show them this figure is how did everyone end up on that block? We can actually go to each home and conduct interviews to find out the history of motions that led each household to this block. Of course, we would want to find out much more. When I began this process, I informally went from household to household to learn the story of each family's lives before they ended up in the neighborhood and since arriving in this neighborhood.

For example, I asked them what they do for a living. My neighbors apparently trusted me more than I expected, as I found out both the legal and illegal ways in which they made money. A couple of the men were sanitation workers; another was a UPS driver. One man was a shrimp boat captain who would go offshore into the Gulf of Mexico for a week or two at a time and then return to the neighborhood (usually bringing back a gift of a bag of freshly caught shrimp for each neighbor). For my immediate Asian Indian neighbors, the husband was an auto mechanic, and the wife worked part time at a clothing store. Another of my immediate neighbors told me they primarily make money selling drugs. I was rather surprised that they would tell me this.

I pressed on, "Could you tell me more?" With little education, the household of four female cousins and their children (it was not clear why no men were there) could not find decent enough employment—housecleaning work mostly—to cover their expenses. They sold drugs around the area and in their home (lots of cars coming and going) to help buy food and send one of the cousins to college. Yes, they thought what they were doing was wrong, and they hoped to get out of the practice, but for now it was the only way they could see to survive.

Also, a mix of people with legal papers and people without papers lived in several of the houses, the latter group earning all of their income "under the table." In fact, a good deal of movement in and out of these homes was the norm—people coming for one month to a year and then leaving. They were part of the continual flow, we found out, between nations. Generally living for free or at little cost in a relative's home, they were there to make money to bring back home, whether to obtain a house back in their home nation or to keep a place they already had.

Another of my neighbors, the Mexican husband and his Puerto Rican wife, were proud of the life they had made. For them, moving to this neighborhood represented success, the realization of a dream of owning a home on a cul-de-sac. The husband, who had at first migrated without papers and with the intent of returning to Mexico, had somehow along the way secured papers, was now legal, and was intent on staying in the United States. He had recently secured a decent-paying blue-collar job at an oil refinery on the east side of town (a thirty-mile commute), and his wife was mostly able to stay home to care for their children. Occasionally she would work odd jobs to make some extra money to have a special birthday party for one of their children or other such special occasion.

Simply asking what people did for a living produced an amazing array of information and understanding. I even went so far as to map their daily movements to and from their places of work, taking children to school, heading to the local Catholic Church or temple. The chart looked like someone had thrown spaghetti on it, with paths going every which way, crisscrossing each other, twisting, and turning.

We could ask so much more to find out about motion—the changing size of their households, for example. As I noted, relatives and friends often came and went from these homes, staying for a few months and then moving on. Tracking where these friends and relatives went and why they came and went would lead us into a whole new realm of motion study.

We should ask other questions: What was life like where they had lived (be that in another country or even just a different part of Houston)? What were their experiences of migration? What challenges did they face in their migration movements? What do they think of their current neighborhood? What are their plans for the future? What do they watch on TV? What do they read? If they worship, where? What do they miss from their earlier life? How often do they travel back to where they are from? What do they take back with them when they do? If they have children, what are their concerns for them?

At one point, I asked my neighbors to take a card with empty homes drawn on them—essentially a figure like Figure 4.1, but with nothing written on it except the empty boxes representing homes—and asked them to fill in their ideal ethnic makeup of a neighborhood. That is, if they could have any ethnic composition of their neighborhood, what would they ideally choose? I did this to understand their preferences, whether their motions that led to living in this neighborhood led to a type of ethnic diversity they were seeking (if they were seeking one at all), and based on this information, to anticipate future body motion.

Many of my neighbors found this a delightful opportunity. What I found was interesting. All of my neighbors wanted ethnic diversity in their ideal neighborhood, but I found two other patterns. First, no matter the level of desired diversity, my neighbors were more likely to place people of their same ethnicity next to them, so the diversity they put down was in the neighborhood, but at the edges. What's more, if my neighbor was right-handed, he or she would put more people of the same ethnicity on the right side of the figure. If they were left-handed, more on the left side. The white families followed these patterns most strongly, and also put down less diversity than others did.

From this and our theoretical understandings, my students and I predicted future migration movement. The whites in the neighborhood would be most likely to move. We also thought that others might consider moving, but that ultimately they would not, given that though they were not always able to realize their ideal neighborhood ethnic composition, the one they were in was likely as close as they could find. We knew that diverse neighborhoods in general are more valued by people of color than by U.S. whites, so we predicted that over time, the neighborhood would stay diverse, but ultimately would lose most whites. In 2004, five years after moving to this neighborhood, my family and I moved to Chicago. In Figure 4.2, I show the ethnic composition of the neighborhood when we left, along with the houses where actual moves occurred and new families moved in. Clearly, our predictions of the change in the white presence in the neighborhood were realized, and even faster than we would have expected. After just five years, all the whites on the block had moved, and in no case was any sold house replaced by a white family. Thus, after five years, the neighborhood remained diverse, with people from across the Americas and elsewhere, but U.S. whites were no longer part of this diversity. This pattern occurred block after block throughout the entire subdivision of about 1,500 homes.

So the concept of motion pushes us further into more studies. What are the implications of these changes for the people in the neighborhood?

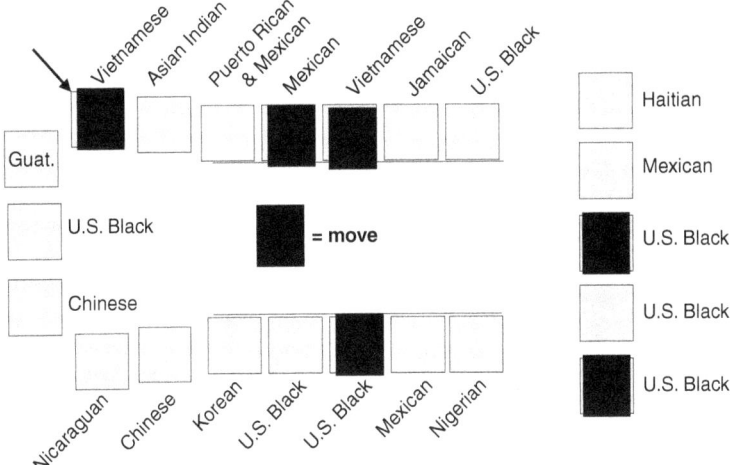

Figure 4.2   Neighborhood in 2004

What does it mean for their thoughts about their futures, for staying or moving on? We could ask so many more questions to learn if anything has changed since ten years earlier.

We could also delve deeply into studying the motions of cultural change for the people in this neighborhood (I do not do so here because our studies in this area are just commencing). How are these people being transformed by their residence here, in a new land or a new side of town? What influence are they having on the community in which they live, and on the communities from which they came? How do they understand themselves? Have their self-identities moved in any way? How about that of their children? Our broader goal: to learn how the Americas are either being tied together or separated by these motions.

Obviously, we could go on, but the point is clear. Armed with the concept of motion, we can learn about the Americas simply by studying a single neighborhood over time. It may be a diverse neighborhood, such as the one being described here, or it may be a neighborhood filled with all Columbian immigrants, for example. Either way, so much understanding about the Americas can be gained by studying it.

What I am suggesting, then, is that as we study and teach the Americas, we can learn much by using the concept of motion, and importantly, by adding to our reading knowledge a practical component to learning— having students actually study a neighborhood or series of them. Whether the discipline is history, religious studies, English, area studies, or a social science, students can benefit from actual hands-on study. And since the concept of motion implies that the diversity is at the doorstep of where we live and is, in fact, realized, especially in the cities of the Americas, there is ample opportunity for such study and for developing our understanding of the richness that is the study of the Americas.

Notes

1. Don Doyle and Marco Pamplona, *Nationalism in the New World* (Athens: University of Georgia Press, 2006).
2. See, for example, Walter Mignolo, *The Darker Side of the Renaissance: Literacy, Territoriality, and Colonization* (University of Michigan Press, 2003).
3. Lester D. Langley, *The Americas in the Modern Age* (New Haven: Yale University Press, 2003), 266.
4. The figures and calculations were made from http://en.wikipedia.org/wiki/Country_population, accessed November 4, 2009. The reported data are current as of 2009, either from actual censuses or estimates extrapolated from the last census and incorporating information about births, deaths, and migration.
5. These claims come from Douglas A. Massey (1999). A recognized expert on international migration, he has spent approximately thirty years building a theory of international immigration.
6. Latapí, Augustín and Susan F. Martin, *Mexico-U.S. Migration Management* (Lanham, MD: Lexington Books, 2008).
7. This figure comes from Jeffrey S. Passel, "Unauthorized Migrants: Numbers and Characteristics," report (Washington, D.C.: Pew Hispanic Center 2005), http://pewhispanic.org/files/reports/46.pdf (January 12, 2010), with unpublished updates from Jeffrey S. Passel. Importantly, these figures include both people who are considered to be in the United States legally and illegally.
8. These figures are taken from http://en.wikipedia.org/wiki/Immigration_to_Mexico (accessed November 10, 2009).
9. From the work of Durand, Jorge, and Douglas S. Massey. "What We Learned from the Mexican Migration Project," in *Crossing the Border: Research from the Mexican Migration Project*, eds. Jorge Durand and Douglas S. Massey (New York: Russell Sage Foundation, 2004), 4.
10. From *Mexico-U.S. Migration Management: A Binational Approach*, eds. Augustín Escobar Latapí and Susan F. Martin, 2008 (Lanham, MD: Lexington Books. 2008), 13.
11. Durand and Massey (2004), 7.
12. Ibid., 10.
13. Sandya Shukla and Heidi Tinsman, eds., *Imagining Our Americas: Toward a Transnational Frame* (Durham: Duke University Press, 2007), 6.

CHAPTER 5

Embodied Meaning: The "Look" and "Location" of Religion in the American Hemisphere

Anthony B. Pinn

The American hemisphere is marked by complex and shifting religious concerns and encounters—resembling a rich mosaic composed of various traditions and orientations, each with its own integrity, history, thought patterns, ritual structures, and frameworks.[1]

For those interested in the nature and meaning of religion within the American hemisphere, particularly as related to theorizing religion, this diversity poses a challenge: Does religion in the American hemisphere possess a fundamental or elemental nature? Is there something that links various traditions across the cultural geographies of the American hemisphere? If so, what is it?

Within this chapter I argue for a fundamental connection, but not one related to ritual structures, doctrinal arrangements, or other external structures of expression. Rather, I argue that the body (as an intertwining of discursive construction and material reality) and how it occupies time and space is the fundamental connection between traditions that make up the rather messy religious geography of the American hemisphere. By this I mean the ways in which bodies—particularly "black bodies"—have been named, discussed, and used within the context of the Americas is a shared experience that cuts across political, social, and cultural differences. I am not arguing that such an understanding of the body is workable in the American context because there is something unique about this context. Instead, I am only arguing that attention to the body offers scholars and students concerned with religion in the

Americas a useful way of mapping religiosity beyond any one particular nation/state. The body so conceived constitutes a shared vehicle of existence that allows for study and reflection across national lines. While the various traditions in the Americas have creedal, ritual, and theological differences, they share concern for and attention to the body. Furthermore, such an understanding of the body and its importance for the study of religion within this hemisphere also calls for attention to a fuller range of source materials. I turn to photographic images because they offer source material that is richly useful in unpacking the nature and meaning of embodied religion in this American hemisphere.

The Black Body and Religion

The manner in which the body is contested and formed suggests, in part, its importance for the study of religion, and I offer the African Diasporic dimension of the American hemisphere as a case study.

Both dominant and presumed docile bodies (labeled "other") take form and function within these processes, and religion, with its concern for the existential and ontological conditions of a group and the layered causes for and resolutions of those conditions offers insights into the nature and meaning of every*body* within the American hemisphere. This is not to say that other expressions and locations of the religious should no longer be played out in the academic study of religion in the American hemisphere. Rather, I suggest that the body is an underexplored and underappreciated framework for such study, although it provides ways to think across and between the differences and similarities between traditions found in this hemisphere. How bodies occupy time and space was a major consideration from the moment Africans were enslaved and held at various points along the African coast, to confinement on slave ships, and to their placement in the intellectual and physical geography of the American hemisphere. And religion marked a way of negotiating the felt and symbolic meaning of space and time.

The religious tone and texture of the American hemisphere holds in tension the love and lure of the black body as cultural construction and as lived reality.[2] The effort of people of African descent to forge life meaning is, in essence, an effort to rescue and redeem their bodies.

Religion, while not the only organizing principle within the American hemisphere, becomes a way of exploring and explaining how bodies—discursive and physical—occupy time and space, and it is an exploration that does not require allegiance to any particular religious doctrine or practice. In this regard, centering investigation on something other than traditional institutional forms, ritual structures, and doctrinal/creedal developments allows religion to be defined in ways not limited to a particular tradition, but rather through a theoretical arrangement that cuts across localized arrangements of the hemisphere. *In this case, religion is understood as the quest for complex subjectivity—the push for greater life meaning. Religion, so understood as the quest for complex subjectivity, is embodied, and the body is the outcome of religion. In this way, efforts to rethink the proper framing and "placing" of bodies as they occupy time and space so as to shift their meaning or importance is not simply a mundane concern; it is religious.*

I am not suggesting that the body provides a firm and solid means by which to access and understand religion. To the contrary, both the discursive and the material body are interpreted with respect to and connected to matters of cultural construction. In this regard, the physical body is not "natural," without "contamination" from language; it is not fixed, and it is not certain.[3] It is present, but never fully captured—framed and informed by language but also constituted out of materiality. The body is constructed, spoken into existing, as a matter of power and knowledge; but it is also more than the "stuff" of language. Hence, modernity as manifest in the American hemisphere involves the construction of discursive bodies and the disciplining of physical bodies through a host of violent and mobile means.[4]

Bodies gather, shape, and encode information about structures of relationship and meaning. This was true within the mechanics of the slave trade—capture and auctions, for example—rebellions and revolutions, through past and more recent strategies of observation, defining, and restriction based on markers such as race and gender. There were numerous and interrelated markers of "fit" within the social body based on the dynamics of the individual's body. For example, "Imperial images of the colonized native American, African, and Asian as eroticized savage or barbarian," Ann Stoler remarks, "saturated the discourses of class."[5] Furthermore, in examining travel accounts, Jennifer Morgan notes the ways in which black female bodies were viewed and depicted so as to maintain racial distinctions and structures of cultural normativity. Their bodies were discursively pulled apart and noted for their reproductive value or, in some cases, their forbidden erotic allure. And their physical bodies were labeled valuable based on their capacity for work. While the expression of preoccupation with black/gendered bodies would change over the course of centuries, what remained in place was an underlying discourse of negative difference based on a continuum of inferiority. It was in counter position to this depiction of black female bodies, as well as depictions of black male bodies, that Europeans constructed themselves.[6] The notions of purity of culture and race undergirding the cataloging of bodies had far-reaching consequences that move between historical moments and that influence response to substantial ontological and existential questions.

The body—discursive and physical—is contested terrain, changing and twisting, but even this says something about the nature and meaning of religion in this "part" of the world.

Picturing the Body and Studying Religion

The manner in which the body occupies time and space has been expressed and questioned by means of the "written text"—for example, in literature such as diaries, slave narratives, plantation records, church records, and so on. Yet, to reshape the study of religion in the American hemisphere in light of embodiment requires a range of source materials. Visual images such as the photograph, when partnered with other source materials such as written texts, offer an important dimension of investigation. While the photographic image does not provide a context for studying religion over the course of the full historical period embedded in the American hemisphere, it does offer useful material by which to interrogate the depiction and discussion of black bodies related to long-standing assumptions and practices.[7] Put differently, although the photographic image is not used during the full period of slavery in the Americas, it does serve to capture assumptions of aesthetic and cultural inferiority used to support the logic justifying slavery. In addition, efforts to resist oppression through an assertion of aesthetic and social value are also marked out through photography. So understood, photography documents long-standing efforts to both justify and critique the oppression of black bodies. For example, prior to the end of slavery in the United States, Harvard University's Louis Agassiz requested images of enslaved Africans from South Carolina to confirm his "scientific" speculation that Africans (and their descendents) were different and inferior by nature. (The images, although of naked black bodies, did not expose scars representing the harsh realities of slavery. Such markings would not have served his purpose, which was to demonstrate physically the difference—and hence, inferiority of black bodies—not to make visible the brutal nature of slavery/discrimination.) He argued that scientific study of their bodies demonstrated this inferiority and that the requested photographs would provide another way to read and affirm this difference; similar photographs were taken of enslaved Africans in Brazil.[8] For some, the initial purpose of the photographs was met. Yet in spite of

Agassiz's intent, those images, having re-emerged in the late twentieth century at Harvard University, portray the "face" of slavery—the way it pulls at the flesh—and the humanity of those black male and female bodies, regardless of efforts to subdue them and mold them into something less significant.[9] These bodies push against a simple read as stigmatized "others," portraying instead a certain type of dignity.

The photographic image, according to Carol Williams, served the development of difference within the context of U.S. expansion by implying boundaries between controlled and uncontrolled geographies, between the corresponding privileged human bodies and assumed inferior bodies. It was used as a way of "capturing" the inferiority of Africans and Indians in North America, and of demonstrating the material value and cultural dominance of Europeans.[10] These photographs, chronicling the grabbing of land in the frontier, served a negative function, in that these images reified the "other" as less aesthetically pleasing, less culturally refined—and available for visual consumption and use by the more "powerful" population. Yet, the photographic image can be depicted as *either* negative or positive. The image speaks twice: the material captured, but also the question of that which the image does not contain. It is this openness, the porous nature of the photographic image, that I invite readers to explore religiously.

There is more than meets the eye in that the language(s) used in the construction of the photographic text invests "every aspect of the photographic space" with a "potential meaning beyond its literal presence in the picture."[11] The public is privatized and the private is made public—the content can be read in a variety of ways for a variety of purposes. A photograph is physically flat but is nonetheless a visual labyrinth, thick in its possible strategies of purpose and layered in its stories of relationships, but always tied to the complexities of embodiment. Although the product of a single "angle" offered by the photographer, the reading of the image as text is not so easily controlled and is tied to the "religious" in part because it nurtures the unanswered questions of existential and ontological importance: What more is here? And who/what am I that I pick from this image/text these certain things? Answering these questions is somewhat archaeological in nature, and what is determined along the way informs structures of the self, the self in relationship to others, and both in relationship to objects. As I have argued elsewhere, the body is suggested as that out of which religion is formed and to what it responds.[12] In this regard, the body can be read and studied as material and discursive evidence of the anxieties, beliefs, practices, and structures of knowledge that informed and shaped the American hemisphere and to which religion responds. Attention to photography promotes an alternate means by which to understand the nature and meaning of embodied religion.[13]

The Photographic Image—Capturing the "Soul"

A simple definition of the photograph is "light-writing," a type of "communication through images."[14] As such, photography changed the manner in which we express and receive information, altering the ways in which we imagine and view others, our surroundings, and ourselves.[15] It represents in content and form the effort to frame the world, to capture moments of experience for consumption and expansion, by offering a "piece of reality."[16] One can make distinctions concerning the use of the photograph as a means of communication, whether the photographs are taken by an expert or amateur. And one can make distinctions concerning equipment used and possibilities regarding mode of presentation. Yet, I am less concerned with debate over photography as "art" or the dynamics of its development, and more concerned with how this "light-writing" is read and what it says about the religious nature and meaning of bodies in time and space.[17] In this regard, I argue that the photograph exposes deep meaning connected to ontological and existential struggles and confessions. "In seeming to capture times and places lost in the past," write Annette Kuhn and Kirsten Emiko McAlliser, "the photograph can disturb the present moment and the contemporary landscape with troubling or nostalgic memories and with forgotten, or all too vividly remembered, histories."[18] Whether of individuals, families, larger units of human relationship, or the environment of which humans are a part, photographs share something of the histories lodged in the open spaces and more hidden corners of the American hemisphere.[19] Each image involves a wrestling over the markers of meaning in ways that pull the viewer in—body and all. In the photographic image, bodies and their surroundings, the signs and symbols associated with them, are held in place but are not rendered stable, and are open to investigation and review.[20] This is an important consideration in that the photographic image, from the time of its introduction in the 1830s in Europe, provided a significant way in which the workings of the American hemisphere were expressed, questioned, and spread.[21]

What the photograph entails can support existing structures—racism, sexism, classism, and so on. It is in this way that the taking of photographs by colonial administrations was a process of surveillance and of documenting the "other" such that typologies of racial differences and cultures of domination were presented and authenticated.[22] And yet, the photograph can resist and challenge the dominance of these structures. Thus, it was also a way by which those "othered" within the context of the Americas could reposition themselves. In addition, these "microhistories" of othered individuals and communities pointed out the problems with the image of the dominant population in the hemisphere, outlining the fragile and fictitious superiority of "white" bodies.[23] For example, tension related to race and cultural identity in locations such as Mexico was displayed and at times contested through images.[24] Through the photographic image, the varied population of peoples presenting diverse mixtures of cultures (i.e., "blood") could be read against colonial processes and legacies of enslavement, and those within the image take on an alternate aesthetic and social meaning/presence. Assumptions and dreams of the inferior "other" are challenged through an altered presentation of the "othered."Thus, bodies depicted could represent the docile and acceptable, or bodies could depict resistance to those standards.[25] The bodies could also both confirm and deny these standards.

Reading photographs is complex and layered in that observation can inform and affirm the fixity of certain bodies. For example, in recent years, the 1990 exhibition "Los Vecinos/The Neighbors" housed at the San Diego Museum of Photographic Arts privileged depictions of Mexicans as problematic bodies marked only by poverty and desperation.[26] The framing of the dominant U.S. cultural and ecopolitical nation was reinscribed through the photographic depiction of bodies. Or, viewing embodiment through embodiment can result in challenges to assumptions of normativity that made slavery reasonable and that continued to buttress the processes of "other-ing" marking the sociopolitical and economic geographies of the Americas. It is through this challenge that structures of life meaning come into view and can be altered to great benefit.[27] As Susan Sontag suggests, photography grabs pieces of the world and places them within our consciousness, allowing us to work (with) them. In this sense, photography became a way for black families of means in various locations in the Americas to "acquire" elements of the world long denied.[28] In this case, familiarity both challenges and affirms normalized bodies and surroundings, and, to the dismay of white supremacy discourses, familiarity points on the marked "sameness" of black and white bodies.

In a related manner, then, turning to images—both photographs and drawings—serves to suggest the manner in which the captured image also demonstrates the attempt to more positively rethink and reframe life and perceptions of othered bodies in (im)proper time and space in various quarters of the Americas. Photography within Latin America, for example, "bore both the promise of modernity as technological advancement and the stain of late-nineteenth and early twentieth-century projects of imperial expansion." In like manner, photographs of the *jíbaro* (rural worker) served to represent at times a symbol of national identity in positive terms through nostalgic adherence to an honorable and ethical structure of relationship to humanity and earth; and at other times, served as a marker of colonialism and racism through assumed connections between colonizers and the much lighter rural workers over and against darker Puerto Ricans. As the United States took control of the island, both interpretations of the image of the rural worker promoted an ethos of "other-ness."[29] At times, the photographic image was used as a way of fixing certain bodies as inferior and reducing identity to natural servility.

Photographic images can be read on one level for what they say about the structures and "look" of imperial domination, but they also offer the viewer material by which to question these workings of "other-ing" within the Americas. Hence, presentation of the body is never a neutral endeavor, but instead is always intentional—layered, complex, and full of possibilities. The images and what they convey is un/natural and un/true. The photograph emerges out of a conflation of forces and interests (i.e., the photographer, those persons photographed, the unspoken stories of the objects, and the larger sociocultural context), and is read in *light* of a set of interests. The photograph as something of an ethnographic process both captures and frees the material body, in part through an opportunity to challenge the truth of discursive construction of certain bodies as inferior in form and content (but not in potentially unproblematic ways), and also by positioning the materiality of the body anew through the "truth" of the image and alternate fields of bodily experience. For example, early photographs commissioned by African Americans in the United States displayed their families as orderly, aesthetically sound, and situated within comfortable surroundings. Depicted in this way, these images signified discourses on black inferiority represented by poor family structure, general antisocial tendencies, and the physical scars of punishment.[30] Images, such as those of strong families, could, as Shawn Michelle Smith remarks regarding images contained in the Paris Exposition of 1900, shift markers of identity such as race "from a single set of cultural practices and progressive potentialities."[31] But these images could also further normalize these same notions of the "proper" role and postures related to the aesthetics of masculinity and femininity for men and women, respectively. This is because images are thought, presented, and read within the larger framework of our cultural world. As Ana Maria Mauad states concerning photographs of certain Brazilian families, photographic images serve as " 'places of memories' that each family developed in forms distinctive to their social situation and historical period."[32]

Fixing of black bodies is not simply a result of white misdeeds. It is also the case that blacks at times fix their own bodies consistent with or as opposition to the status quo, but with what we might hold as questionable outcomes. Or, photography could promote counternarratives whereby black bodies resist negative depiction and inferior framings.[33] Concerning the latter point, from portrayals of slavery's brutal effects on the human body to the popular images of the civil rights struggle, the photographed image has shaped some of the content and form of resistance. These images, like so many others, are fixed in time and space, but they are also *moving* signs and gestures open to a variety of readings and corresponding meanings. The lessons learned through these images are not of necessity long-lasting; however, they do provide at least a moment of dissonance during which those who have been other-ed can assert themselves anew. And religion becomes the cartography of this assertion. Again, religion involves an embodied quest for complex subjectivity, and photography provides a source material for interrogating religion so conceived.

Undergirding the impact of black bodies in the United States, for example, was the manner in which certain images were meant to suggest Christian bodies rightly positioned in family and society through proper relationship to God. In this way, images of black families involved nonwritten affirmation of theories of domesticity advocated by prominent ministers and embraced, at times in modified form, by numerous black women and men within church circles.[34] The women in the photographs were not presented as "mammies" in domestic servitude to white children, but instead the signs point to their focused care on their children— their families. Domesticity as the proper station of the wife remained a patriarchal fetter, but in these images, the women were tied to their own homes and were not items of a sidebar quality. In these black family photographs, the women and children are not shadowed by the threat of danger from white supremacy, but rather are protected and cared for within the context of a nurturing domestic cocoon.[35] Worlds of difference presented visually implode, as the private arrangements of black life that run contrary to the discourse of inferiority are made public.

Images—that is, bodies in pose—matter. And through the photographic image we capture something of the human's ability to reshape substances for new uses, giving them new meaning. Photographs of "founding" pastors and key leaders of churches displayed, for example, in the vestibule of church buildings say something about the socio-religious and ethical commitments of particular congregations. This is because the images of exemplary leadership are meant to present visual history but also shadow the ongoing concerns and functions of the church. Placement of these images is a statement of congregational identity. One can read a similar visual lesson in the images of churchwomen in Sunday hats and corresponding outfits wherein the aesthetic commitments of the congregation are outlined as an indicator of how one prepares the body for worship as a community activity.[36]

These examples are not meant to suggest that the significance of photography in relationship to the study of religion in the American hemisphere is dependent on overt religious signs and symbols marking the image.[37] To the contrary, the importance of photographs for the study of religion in the American hemisphere is not premised on the value of the image based on familiar markers of the religious, but rather, on the more general ways in which the arrangement and presentation of bodies speaks to the struggle for complex subjectivity.[38] Neither the images mentioned earlier, nor Renée Cox's image of the Christ figure ("It Shall Be Named") and the earlier photographs of the converted "others" used by Christian missionaries to document their North American efforts represent more fertile source material for theologizing than, say, home photographs. These and countless other images teach lessons, convey information, offer warnings, and in a general sense shape the content and form of the discourses guiding individual and collective life. What they provide is not limited to the first circle of connection (e.g., those who commissioned or took the photographs); they are also of value with respect to the everyday "voices" beyond that first circle.

The photographic images produced are layered and thick with meaning(s). By implication, the need or desire to capture a moment in time through photographs invests that particular moment and its contents with worth.[39] Even if in another context—unnoticed by the camera—a similar moment is indistinguishable based on significance, once captured in a photo, viewers assume it must have some type of value, even if this value is not immediately apparent.[40] The thought behind the taking of the image, the arrangement of the body—the pose—and the "dressing" of the body, as well as how the actual photograph is handled and shared, say something about the structuring of identity. These say something about the scope and importance given to the ways in which bodies occupy time and space. To borrow from Theophus Smith, photographic images became a process of conjuration, a means of "conjuring culture," whereby lived experience is interrogated and reshaped.[41]

Religious usefulness and importance are not premised on how well photographs correspond to standard signs and symbols.[42] Clearly, interpreting or reading photographs poses a challenge, but it is no greater than the challenge present in the "reading" of any source material in that all such materials are marked by the dilemma of shifting contexts and the fragile nature of historical/cultural memory. Such complications are unavoidable and should be recognized as an inherent "risk" when studying religion within such a messy geography.

Photographic images as a general product have value for the study of religion in that photographic images speak to modalities of meaning and subjectivity within the context of the individual and the individual in relationship to other persons as well as the "environment" (both natural and constructed). Keep in mind how religion was defined earlier in this chapter: Religion is the quest for complex subjectivity, and is representative of the yearning for life meaning. So defined, the religious is a certain framing of human experience, not a unique modality of experience. Stepping outside the familiar framing of the religious, one finds something of value and insight in the more general experiencing of life. This is because the embodied quest for complex subjectivity—that is, religion—is expressed in complex and significant ways through a variety of life situations and experiences. How bodies occupy time and space is the "stuff" of religion, and embodied theological thought maps describe the form and meaning of this occupation. The photograph captures this occupation of time and space, and makes it available for analysis. And whereas certain modalities of body placement as religious are unavailable to all—such as spirit possession—the photograph is not of necessity so rarefied or specialized. The photograph renders things assessable.

Photographic Determinations

The turn to the visual image avoids the tendency to speak in terms of a generic or "ideal" black person by forcing attention to particular human bodies—not abstractions, but rather the movement and form of particular fleshy bodies. But it does so in a way (in light of the historical context captured) that also notes the social nature of the body—embodiment as reinforcement of status quo or a challenge to the same. The aim should be to recognize the merit of lived experience, that is, the body as material in time and space. Avoidance of a particular interpretation as authoritative is required.

While having selected the particular images used in study of religion so conceived suggests a certain type of control on the part of the scholar, the aim should be to acknowledge the implications of that unavoidable arrangement, and yet to do one's best to limit any additional (and avoidable) exercise of power over the reader's study of the images as religiously significant. What Williams argues concerning early images of Native Americans, I believe, holds true concerning marginalized groups in more general terms: Captions amplify a particular meaning, a particular "framing" of the image and work to limit other possible interpretations.[43] Furthermore, I disagree with arguments that photographic images, unlike written texts such as novels, lack attention to integrity, character, and so on. I do not find convincing arguments that photographs lack context—beginning and ends—thus making them poor depictions of "struggle," based on the impact, for example, of photographs related to the Civil Rights Movement. *The goal is to highlight the body as flesh, as materiality (as "text"), and from that recognition to then study religion.*[44]

Photographs hide and expose; they challenge our perception of things and their placement. As Susan Sontag observes, "in teaching us a new visual code, photographs alter and enlarge our notions of what is worth looking at and what we have a right to observe."[45] Hence, we read through and dig into photography/photographs (of both a personal and a public nature) for what they can tell us about the "stuff" of embodied life. Religious concerns and categories are presented and challenged through the photographic material available to us. To the extent that religious thinking and experience involve the description and analysis of bodies occupying time and space, and the significance/content of that occupation, photographs provide a potentially powerful source material. While photographs do not stand alone as sources of religious knowledge and experience, and must be combined with other "texts" that say something to and about the sociopolitical and culture context embedded in the image, they do provide useful materials that expose something of the nature and meaning of religion within the American hemisphere.

Notes

1. One of the early and most widely discussed works related to the cultural exchange marking this hemisphere is discussed by Paul Gilroy in terms of the "Black Atlantic." See Gilroy, *The Black Atlantic: Modernity and Double Consciousness* (Cambridge: Harvard University Press, 1993). For challenges to Gilroy's conceptualization based on its relative exclusion of Africa and Asia, see Tiffany Ruby Patterson and Robin D. G. Kelley, "Unfinished Migrations: Reflections on the African Diaspora and the Making of the

Modern World," *African Studies Review*, Vol. 43, No. 1 (Special Issue on the Diaspora, April 2000): 11–45; Michael Gomez, ed. *Diasporic Africa: A Reader* (New York: New York University Press, 2006).

2. See Foucault, *The History of Sexuality*, vol. 1–3 (New York: Vintage Books, 1978–1986); Foucault, *Discipline and Punish: The Birth of the Prison* (New York: Vintage Books, 1979); Susan Bordo, *Unbearable Weight: Feminism, Western Culture, and the Body* (Berkeley: University of California Press, 1993); Chris Shilling, *The Body and Social Theory*, 2nd ed. (London: Sage Publications, 2003); Bryan S. Turner, *The Body and Society: Explorations in Social Theory* (New York: Basil Blackwell, Inc., 1984).

3. I make this qualification here in part as a response to theological perspectives on the body offered by theologians such as Sarah Coakley. See Sarah Coakley, "Introduction," in *Religion and the Body*, Coakley, ed. (New York: Cambridge University Press, 1997), 2–10.

4. For an interesting study of racial formation within the context of globalization, see Kamari Maxine Clarke and Deborah A. Thomas, eds. *Globalization and Race* (Durham: Duke University Press, 2006).

5. Laura Ann Stoler, *Race and the Education of Desire: Foucault's History of Sexuality and the Colonial Order of Things* (Durham: Duke University Press, 1995), 124.

6. Jennifer L. Morgan, "Male Travelers, Female Bodies, and the Gendering of Racial Ideology, 1500–1700," in *Bodies in Contact*, Tony Ballantyne and Antoinette Burton, eds. (Durham: Duke University Press, 2005), 55.

7. Mitchell Stephens, *The Rise of the Image and the Fall of the Word* (New York: Oxford University Press, 1998), 58–69.

8. My telling of this story is based on Alan Trachtenberg, *Reading American Photographs: Images as History, Mathew Brady to Walker Evans* (New York: Hill and Wang, 1989), 53–54; Eleanor M. Hight and Gary D. Sampson. "Introduction: Photography, 'Race,' and Post-Colonial Theory," in *Colonialist Photography: Imag(in)ing Race and Place*, Hight and Sampson, eds. (New York: Routledge, 2002.), 3.

9. Readers might be interested in comparing the intent behind these depictions of enslaved Africans with the use of photography to portray privileged womanhood. The J. T. Zealy photographs of enslaved Africans are available in Alan Trachtenberg, *Reading American Photographs: Images as History, Mathew Brady to Walker Evans* (New York: Hill and Wang, 1989), 55. Also see Carol J. Williams, *Framing the West: Race, Gender, and the Photographic Frontier in the Pacific Northwest* (New York: Oxford University Press, 2003); Shawn Michelle Smith, *American Archives: Gender, Race, and Class in Visual Culture* (Princeton: Princeton University Press, 1999).

10. Carol J. Williams, *Framing the West: Race, Gender, and the Photographic Frontier in the Pacific Northwest* (New York: Oxford University Press, 2003), 4–31.

11. Clarke, *The Photograph: A Visual and Cultural History* (New York: Oxford University Press, 1997), 33.
12. See, for example, Anthony Pinn, *Embodiment and the New Shape of Black Theological Thought* (New York: New York University Press, 2010); Pinn, "Introduction" and "A Beautiful *Be-ing*: Religious Humanism and the Aesthetics of a New Salvation," in Pinn, ed. *Black Religion and Aesthetics: Religious Thought and Life in Africa and the African Diaspora* (New York: Palgrave Macmillan, 2009); Pinn, *Terror and Triumph: The Nature of Black Religion* (Minneapolis: Fortress Press, 2003).
13. Scott McQuire, *Vision of Modernity: Representation, Memory, Time and Space in the Age of the Camera* (Thousand Oaks, CA: Sage Publications, 1998), 7.
14. Graham Clarke, *The Photograph*, 11; Alan Trachtenberg, *Reading American Photographs: Images as History, Mathew Brady to Walker Evans* (New York: Hill and Wang, 1989), 4. I agree with Clarke that photographs are read, suggesting through this language an active process (Clarke, 27).
15. Photography is widely used in other disciplines, such as history, anthropology, and sociology. However, scholars of the study of religion in the African Diaspora have been somewhat slow to move in this direction.
16. Bianca Stigter, "Mirrored Images: The World Reflected in Photographed Eyes," in Frank van der Stok, Frits Gierstbergy and Flip Bool, eds. *Questioning History: Imagining the Past in Contemporary Art* (Rotterdam: NAi Publishers, 2008), 16.
17. Miles Orvell, *American Photography* (New York: Oxford University Press, 2003), 141–161.
18. Annette Kuhn and Kirsten Emiko McAllister, "Locating Memory: Photographic Acts—An Introduction," in Kuhn and McAllister, eds. *Locating Memory: Photographic Acts* (New York: Berghahn Books, 2006), 1.
19. See Martha Langford's work on the photographic album: Langford, "Speaking the Album: An Application of the Oral-Photographic Framework," in Kuhn and McAllister, eds. *Locating Memory: Photographic Acts* (New York: Berghahn Books, 2006), 223–246.
20. Andrew Quick, "The Space Between: Photography and the Time of Forgetting in the Work of Willie Doherty," in Kuhn and McAllister, eds. *Locating Memory: Photographic Acts* (New York: Berghahn Books, 2006), 162–171.
21. Paul S. Landau, "Empires of the Visual: Photography and Colonial Administration in Africa," in Landau and Deborah D. Kaspin, ed. *Images and Empires: Visuality in Colonial and Postcolonial Africa* (Berkeley: University of California Press, 2002), 141; Esther Gabara, *Errant Modernism: The Ethos of Photography in Mexico and Brazil* (Durham: Duke University Press, 2008), 13.
22. Annette Kuhn and Kirsten Emiko McAllister, "Locating Memory: Photographic Acts—An Introduction," in Kuhn and McAllister, eds.

Locating Memory: Photographic Acts (New York: Berghahn Books, 2006), 6; Eleanor M. Hight and Gary D. Sampson, "Introduction," *Colonialist Photography*, 2–3.
23. By microhistories is meant focused attention on a particular event, story, and so on. In this case, it references focused attention on the content of particular photographic images. For information on this approach, see, for example, James F. Brooks, Christopher R. N. DeCorse, and John Walton, eds. *Small Worlds: Method, Meaning, and Narrative in Microhistory* (Santa Fe: School for Advanced Research Press, 2008).
24. Gabara, *Errant Modernism*, 177–186.
25. Landscapes and other images that do not privilege the human presence can also serve to enhance black theological thought by forcing analysis of a complex range of relationships and notions of subjectivity that bring human bodies into contact with a wider range of other modes of life.
26. Aida Mancillas, Ruth Wallen, and Marguerite R. Waller, "Making Art, Making Citizens: Las Comadres and Postnational Aesthetics," in Lisa Bloom, ed. *With Other Eyes: Looking at Race and Gender in Visual Culture* (Minneapolis: University of Minnesota Press, 1999), 107–108.
27. Bloom, "Introducing With Other Eyes," in *With Other Eyes*, Bloom, ed. (Minneapolis: University of Minnesota Press, 1999), 3.
28. Susan Sontag, *On Photography* (New York: Picador, 1977), 4.
29. Gabara, *Errant Modernism*, 3–4; Oscar E. Vázquez, "'A Better Place to Live': Government Agency Photography and the Transformation of the Puerto Rican Jíbaro," in Hight and Sampson, 288–290.
30. Liam Buckley, "Studio Photography and the Aesthetics of Citizenship in The Gambia, West Africa," in Edwards, Gosden, Phillips, eds. *Sensible Objects* (Oxford: Oxford University Press, 2006), 61–64.
31. Shawn Michelle Smith, *American Archives: Gender, Race, and Class in Visual Culture* (Princeton: Princeton University Press, 1999), 161. The mere fact that a black family (or individual) could afford to make use of photography spoke to economic standing and social positioning unavailable to many.
32. Ana Maria Mauad, "Composite Past: Photography and Family Memories in Brazil (1850–1950)," in Richard Cándida Smith, ed. *Art and the Performance of Memory: Sounds and Gestures of Recollection* (New York: Routledge, 2002), 215.
33. Simon J. Williams and Gillian Bendelow, *The Lived Body: Sociological Themes, Embodied Issues* (New York: Routledge, 1998), 190; Smith, *Photography on the Color Line*, 158; Allan Sekula, "The Body and the Archive," in Mariam Fraser and Monica Greco, eds. *The Body: A Reader* (New York: Routledge, 2005), 164; Hazel V. Carby, *Race Men* (Cambridge: Harvard University Press, 1998).
34. For a discussion of domesticity frameworks within the oldest black denomination in the United States, see Julius H. Bailey, *Around the Family*

Altar: Domesticity in the African Methodist Episcopal Church, 1865–1900 (Gainesville: The University Press of Florida, 2005).

35. For example, see the photo "A Hampton Graduate at Home." It is available in a number of books, including James Guimond, *American Photography and the American Dream* (Chapel Hill: University of North Carolina, 1991), 34.
36. Michael Cunningham, *Crowns: Portraits of Black Women in Church Hats* (New York: Doubleday, 2000). Numerous websites attest to the importance of hats as a component of church attire. Examples include "Harlem's Heaven Hat Boutique," http://www.harlemsheaven.com/church-hats and EssenceHat.com http://www.essencehat.com/ (accessed July 5, 2009).
37. In terms of photographs with an overtly religious imagery, readers might consider the F. Holland Day portraits dealing with the last words of Christ (1898). For information on Day, see, for instance, Patricia J. Fanning, *Through an Uncommon Lens: The Life and Photography of F. Holland Day* (Amherst: University of Massachusetts Press, 2008).
38. See David Morgan for an insightful discussion on Christian devotion based on popular images of Jesus Christ. Morgan, *Visual Piety: A History and Theory of Popular Religious Images* (Berkeley: University of California Press, 1999).
39. Susan Sontag, *On Photography*, 28.
40. Some of this was made clear to me in graduate courses with Dr. Richard R. Niebhur, whose interest in works such as *Let Us Now Praise Famous Men* (James Agee and Walker Evans [New York: Mariner Books, 2001]) exposed students such as myself to the rich religious/theological significance of everyday materials.
41. Theophus Smith, *Conjuring Culture: Biblical Formations of Black America* (New York: Oxford University Press, 1994).
42. "It Shall Be Named" can be viewed at www.reneecox.net/series02/series02_11.html. Carol J. Williams discusses the use of photographic images of Native Americans as part of sociopolitical and religious expansion in North America. See Williams, *Framing the West: Race, Gender, and the Photographic Frontier in the Pacific Northwest* (New York: Oxford University Press, 2003), 12–17, 85–107.
43. Williams, *Framing the West*, 27–28.
44. See Mitchell Stephens, *The Rise of the Image and the Fall of the Word* (New York: Oxford University Press, 1998), chapter 6. Also see Susan Sontag's *On Photography* (New York: Picador, 1977); Pinn, "King and the Civil Rights Movement: Thoughts on the Aesthetics of Social Transformation," in Timothy Jackson and Robert Franklin, *Cambridge Companion to King*, forthcoming.
45. Susan Sontag, *On Photography*, 3.

CHAPTER 6

Primeval Whiteness: White Supremacists, (Latin) American History, and the Trans-American Challenge to Critical Race Studies

Ruth Hill

In both the United States and Latin America, racial myths projected onto the colonial period still masquerade as models for talking and teaching about race in the Americas. Among the most tenacious of such myths are the racial hybridity paradigm attached to colonial Iberian Atlantic history and its mirror image: British Atlantic history and the racial purity paradigm. This hybridity-purity dyad has dominated mainstream historiography and the public sphere in the Americas, on the left and the right, since the nineteenth century, and continues to exercise an enormous influence over scholars and students in both Americas.[1] Indeed, just as religious conservatives and white supremacists in the United States sometimes find themselves on common ground, liberal scholars of race and white supremacists share an attachment to those twin paradigms that structure their interpretation of the histories of the United States and Latin America.[2]

The assumption that racial mixture was the norm in colonial Spanish America and racial integrity the norm in colonial Anglo-America, is linked within white supremacist doctrine to another myth about the racial origins of the Americas: the Aztecs, Incas, and Mayas as Caucasians/Aryans/Whites. Since the heyday of white supremacism in the early twentieth-century United States (and in Latin America, where it often donned the ostensibly liberal disguises of *mestizaje* and *indigenismo*), the myth of a primeval or pre-Columbian whiteness—of a white Aztlan—has been constructed from eugenics and other investigations into race and the history of Western and non-Western peoples. White supremacist interpretations of the history and prehistory of the Americas have proven to be as resilient as they are reductive, and digital technologies are expanding the reach of racial myths about the pre-Columbian Americas and about the present-day United States and Latin America. Today, white supremacist groups such as Aryan Nations and Stormfront are breathing new life into the myth of white Aztlan.

Affinities between analog and digital white supremacists are many: their recalcitrant desire to embrace new technologies and methodologies in the pursuit of their cause; their overlapping hatred for Africans, African Americans, Latin Americans, and Latino-Americans; and their unshakeable belief in the Spanish and Portuguese as chosen (white) peoples who forfeited first their racial integrity, and then their imperial power. Nonetheless, white supremacism has never been monolithic, and white supremacist interpretations of the prehistory and history of the Americas should be studied and taught in a manner that respects the material differences between white supremacists from the early twentieth century and contemporary white supremacists. Doing so reveals a great deal about white supremacism in the United States, bringing to the fore its national and hemispheric itineraries, and about how we talk—and do not talk—about race in the college classroom. Moreover, because they are trans-American in and of themselves, white supremacist accounts of the racial history of the Americas challenge scholars and students working in critical race studies to think and respond comparatively. As a result, Latin Americanists and Americanists—and our students, who will one day become scholars—are encouraged to dialogue with each other about issues ranging from the new media to immigration, religion, and politics.

Long before the Critical Race Theory (CRT) movement was launched by a handful of professors at U.S. law schools in the mid-1980s, race was studied, theorized, and taught to thousands of undergraduate students at U.S. universities across different disciplines, departments, and programs.[3] There can be no doubt, however, that CRT has spurred critical race studies to become more sophisticated, theoretically, and more engaged (at least rhetorically) with power issues within and without the classroom. Far too little, however, has been done within critical race studies to problematize and conceptualize race in Latin America, and even less has been done to understand race in the Americas—that is, race as a floating signifier that means different things to different people, in the same place or in different places, at the very same time. As a nascent subfield of critical race studies, comparative racial and ethnic studies focus primarily on race and ethnicity in the United States, only rarely venturing across the mental and physical borders that divide what were commonly known in the nineteenth century as "North America" and "South America." In Latin America itself, race is often denied any ontological status whatsoever, and wherever comparative racial and ethnic studies are attempted, they tend to focus on different races or ethnicities in a specific country or area of Latin America. Academic discussions around white, black, brown, and yellow—with all of their myriad definitional complications—and reductive paradigms in critical race studies of the Americas, which blur what Stuart Hall has termed the "historical specificity of race," are both related to white supremacism in surprising, and significant, ways.

The hemispheric narrative of race in the colonial Americas that ascribes purity and hybridity to the British Atlantic and Iberian Atlantic, respectively, was largely written by travelers, historians, and polemicists in the nineteenth century.[4] Massive immigration to the United States and Latin America throughout that century had somehow to be accommodated within that narrative. In the nineteenth century, the Anglo-American melting pot had a Latin American counterpart: *mestizaje*. The ideology of *mestizaje*, like that of the melting pot, was largely driven by the conviction that biological and/or cultural whitening would redeem some races—not all races, but enough members of some races to build a nation around.[5] Human variation was not, of course, an exclusively hemispheric concern: It had become a global preoccupation. The discourse of degeneration in nineteenth-century and early twentieth-century Europe was punctuated by phrases such as "Anglo-Saxon Superiority," "Latin decadence," "Rising East," and "the future of the White Race."[6] That discourse of decline echoed throughout Latin America, provoking an array of responses.[7] In the early twentieth-century United States, key metaphors appeared to signal that the organizational matrix of that discourse of decline had shifted from geography and what we today call ethnicity to folk race, or color. The Yellow Peril, Black Scare, rising tide of color, and so on were all about color, the color line, and what nativists viewed as the multicolored invasion of America. From the second half of the nineteenth century forward, there was an *expansion* of whiteness in the United States through the "melting" of different white races, but both the disappearance of the mulatto category and the appearance of the Mexican category in the 1930 U.S. Census suggested that color, geography, and ethnicity would not be easily parted.[8]

For early twentieth-century white supremacists in the United States, the primary concern was reinforcing the "Color Line" by restricting immigration from Spain, Italy, Portugal, Africa, Asia, and Latin America, and by maintaining the separation of the races in the United States, particularly in the South, where the majority of African Americans were living in the early twentieth century. The seat of white supremacist ideology was in the northeastern United States: Madison Grant and Lothrop Stoddard were eugenicist clarions of the anti-integration and nativist movements. In Grant's *The Passing of the Great Race or the Racial Basis of European History* (1916) and *The Conquest of a Continent or the Expansion of Races in America* (1933), and in Stoddard's *Rising Tide of Color Against White World-Supremacy* (1921) and *Clashing Tides of Colour* (1935), the goal was not simply to apotheosize whiteness and Aryan/Caucasian/European civilization, but to establish that whiteness had distinct contours—that it could be mapped.[9]

Though they achieved an international following, neither Grant nor Stoddard had formal training or expertise in race matters. Grant was a wealthy wild game hunter and conservationist who served as chairman of the New York Zoological Society, regent of the American Museum of Natural History, and councilor of the American Geographical Society. The 1926 French translation of his *Passing of the Great Race* was prefaced by Vacher de Lapouge, an inspiration for the German architects of Nazi racial policy.[10] Stoddard was a Harvard man. His *Rising Tide of Color* was prefaced by Grant and would lead to a personal meeting with Adolf Hitler, chronicled in *Into the Darkness: Nazi Germany Today*.[11] Neither Grant nor Stoddard was an original thinker, and yet their works constituted what could be called coffee-table eugenics: Their arguments shaped public discourse and informal conversations and prognostications on race in the United States and beyond. Such was their national and international celebrity, in fact, that scientists who worked on race, heredity, and genetics would feel pressed to respond to their claims for several decades.[12]

White supremacists were determined to convince Americans to abandon blending or fusion (*mestizaje*, or the melting pot) as a racial paradigm and national ideal. In *Passing of the Great Race*, Grant set his sights on educating whites (subdivided into Mediterraneans, Alpines, and Nordics) on protecting and preserving their racial heritage and integrity. Grant was panicked because the skulls of Jews and Italians had not grown any larger after they came to the United States, though an unnamed anthropologist (probably Franz Boas) had instructed the federal government that the American environment would alter these "inferior races." Shortly after ridiculing the influence of environment

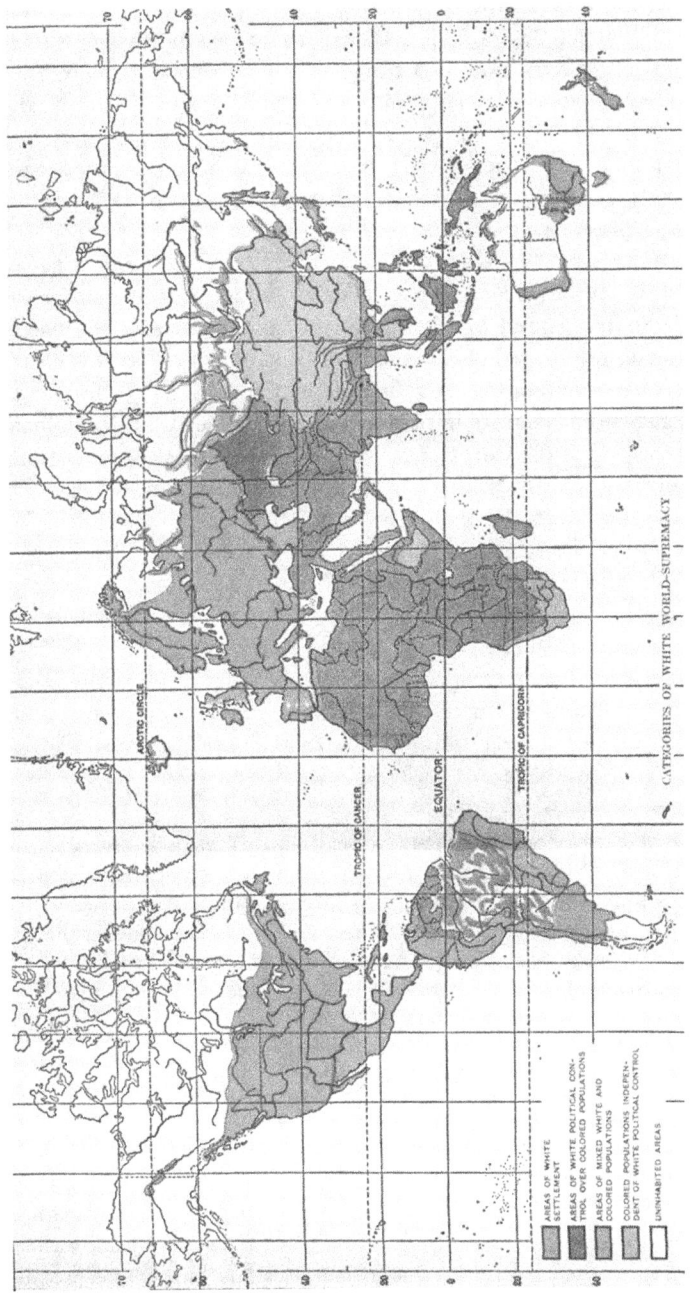

Figure 6.1 Lothrop Stoddard. *Rising Tide of Color Against White World-Supremacy* (1921)
Source: Public domain.

on race, Grant impales Mexico on the melting pot paradigm: "What the Melting Pot actually does in practice, can be seen in Mexico, where the absorption of the blood of the original Spanish conquerors by the native Indian population has produced the racial mixture which we call Mexican, and which is now engaged in demonstrating its incapacity for self-government."[13] At the opening of the twentieth century, *mestizaje* in Mexico represented for Grant the inevitable outcome of the melting pot in the United States.

A few years later, in "Alien Invasion," a chapter from *Conquest of a Continent*, Grant railed against the "new immigration" from Europe—primarily Spaniards and Southern Italians, and Jews from the Near East. In another chapter, "Our Neighbors on the South," it was clear that Mexican immigration was *Indian* immigration for Grant, and nothing good could come of it. Still, his map of Mexicans in the United States according to the U.S. Census of 1930 shows their concentration in the Southwest, with little or no presence elsewhere, and Grant was cheered by the Great Depression because it had encouraged them to return home. In keeping with the intense negrophobia of first-generation white supremacists, Grant's foremost fear was of black immigration, especially from the West Indies, "a standing menace to the United States immigration policy."[14]

For his equally apocalyptic colleague Stoddard, oligarchy and "Aryanization through wholesale European immigration" were two possible routes to stabilization in the Southern Cone (Argentina, Chile, and Uruguay), but the rest of Latin America was doomed:

> Persons of unmixed Spanish or Portuguese descent are relatively few, most of the so-called "whites" being really *near*-whites, more or less deeply tinged with colored bloods. It is a striking token of white race-prestige that these near-whites, despite their degeneracy and inefficiency, are yet the dominant element; occupying, in fact, much the same status as the aristocratic Creoles immediately after the War of Independence. Nevertheless, the near-whites' supremacy is now threatened. Every decade of chronic anarchy favors the darker half-breeds, while below these, in turn, the Indian and negro full-bloods are beginning to stir, as in Mexico.[15]

Undergraduate students readily observe that Stoddard lumped different peoples of color together ("Indian and negro full-bloods" in the previous passage), whereas the curricula at most colleges and universities do not permit students to study race in a trans-American fashion that might bring to the surface the shared histories and burdens of racism

Primeval Whiteness • 115

in the United States and Latin America. Though Stoddard's recourse to color-coded maps conjures up an ironclad racial geometry, negrophobia and Latinophobia slip into one another, and into the Yellow Peril, as the *Rising Tide of Color* advances.[16] Tropical Latin America, or "mongrel America," was imperiled by the rise of yellow peoples, he believed:

> There is [...] another dynamic which might transform mongrel America. This dynamic is yellow Asia. It thrills to novel ambitions and desires. Avid with the urge of swarming myriads, it hungrily seeks outlets for its superabundant vitality. We have already seen how the Mongolian has earmarked the whole Far East for his own, and in subsequent pages we shall see how he also beats relentlessly against the white world's race-frontiers. But Mongrel America! What other field offers such tempting possibilities for Mongolian race-expansion? Vast regions of incalculable, unexploited wealth, sparsely inhabited by stagnant populations cursed with anarchy and feeble from miscegenation—how could such lands resist the onslaught of tenacious and indomitable millions?[17]

While encouraging students to reflect on the overlappings of races, colors, and racisms in Stoddard's polemic, it is essential to pause on the Yellow Peril and a chunk of its legacy to the United States in the twentieth and twenty-first centuries: the model minority and perpetual foreigner myths that students can read about in Frank Wu's *Yellow*.[18] Stoddard feared the Japanese and Chinese peoples because he was convinced that they were sophisticated and similar to white peoples in their abilities:

> White men must get out of their heads the idea that Asiatics are necessarily "inferior" [:] while Asiatics do not seem to possess that sustained constructive power with which the whites, particularly the Nordics, are endowed, the browns and yellows are yet gifted peoples who have profoundly influenced human progress in the past and who undoubtedly will contribute much to world-civilization. The Asiatics have by their own efforts built up admirable cultures rooted in remote antiquity and worthy of all respect. They are to-day once more displaying their innate capacity by not merely adopting, but adapting white ideas and methods.[19]

Stoddard envisioned Asian incursions into not only "mongrel America" but also sub-Saharan Africa:

> To permit Asiatic colonization and ultimate control of these vast territories with their incalculable resources would be to overturn in favor of Asia the political, the economic, and eventually the racial balance of

power in the world. [...] We must resolutely oppose both Asiatic permeation of white race-areas and Asiatic inundation of those non-white, but equally non-Asiatic, regions inhabited by the really inferior races.[20]

Here the Yellow Peril and the Black Scare converge to reaffirm Stoddard's unbridled anxieties about the white man's loss of "colonization and ultimate control" over the nonwhite territories depicted in his racial maps—and, ultimately, over the United States.

It is very probable that Stoddard's views on Asia and Latin America in *Rising Tide of Color*, like those of Grant in *Passing of the Great Race* and *Conquest of America*, were shaped by a long-forgotten study by Francisco García Calderón, *Les Democraties Latines* (1912), published in English translation as *Latin America and Its Progress* (1913).[21] The author was the son of the former Peruvian President Francisco Calderón; the book's preface was written by the future French President Raymond Poincaré. Both Grant and Stoddard were living in Manhattan when the *New York Times* published Edward Bradford's gushing review that highlighted García Calderón's views on race and class in the Americas and his fear of imperialism stemming from the United States, Germany, and Japan.

Degeneration was not an exclusively European disorder, to tilt Daniel Pick's title. García Calderón devoted special attention to New York City in his account, and several years before Grant and Stoddard, he warned of the dangers of the melting pot and class divisions in the United States. Though no direct evidence links García Calderón to digital white supremacists today, the fears and lamentations expressed by him nearly a century ago run rampant on white supremacist websites in blogs and editorials on race, immigration, and America's alleged abandonment of God and white heritage. One hundred years from his present, America as the Peruvian historian and diplomat knew it would no longer exist:

> An octopus of a city, New York might be taken as the symbol of this extraordinary Nation; it displays the vertigo, the audacity, and all the lack of proportion that characterizes American life. Near the poverty of the Ghetto and the disturbing spectacle of Chinatown you may admire the wealth of Fifth Avenue and the marble palaces which plagiarize the architecture of the Tuscan cities. Opposite the obscure crowds of immigrants herded in the docks you will see the refined luxury of the plutocratic hotels, and facing the majestic buildings of Broadway the houses of the parallel avenues, which are like the temporary booths of a provincial fair. Confusion, uproar, instability—these are the striking characteristics of the North American democracy. [...] In a hundred years

men will seek in vain for "the American soul," the "genius of America," elsewhere than in the undisciplined force or the violence which ignores moral laws.[22]

On race in colonial times, García Calderon was clear that miscegenation had doomed the future Latin American republics, just as Grant and Stoddard would argue a few years later, as if they had discovered the racial key to understanding the differing histories of the Americas. The Peruvian's portrait of colonial Spanish America underscores the racial chaos of the time and place, contrasting it with the taut Color Line that allegedly divided colonial Anglo America. Bradford summed up García Calderón's subscription to the purity-hybridity racial dyad to explain the differences between the United States and Latin America:

> England and Spain alike exploited their colonies, and the colonies revolted against both. But racial characteristics modified the developments from this momentous change, in itself so similar in each case. In the case of the Anglo-Saxon the climate alone modified the race. Both Dutch and English held themselves sternly aloof from the aborigines, the Indians, forcing them West and exterminating them. The race was preserved in its purity. In South America conquerors and conquered intermingled. Spanish and Portuguese took Indian wives or women; the leaders married princesses of Mexico or Peru; the soldiers founded provisional homes in the colonies they guarded. The half-castes became the masters by force of numbers, conceiving a thirst for power and the hatred of the proud and overbearing Spaniards and Portuguese [...]. Wars broke out between the Iberians and the Americans, the half-castes of more crosses than could be calculated, and of names which are strange to the simple classifications of miscegenation of whites and blacks, and already forgotten with the end of slavery among us. [...] Grotesque generations with every shade of complexion and every conformation of skull were born in America from unions stimulated by the kings of Spain.[23]

The Peruvian's readers were supposed to infer (as Grant and Stoddard evidently did) that miscegenation is wrong not only aesthetically and morally, but also politically, because it creates a preponderance of mixed-race peoples who overthrow their imperial (white) masters. Having students read excerpts from García Calderón (or from Bradford's review) alongside excerpts from Grant's and Stoddard's works to debunk the myth that Latin Americans have always been "more liberal" about race, that is, more accommodating of somatic differences because of their alleged commitment to the racial hybridity model.

That myth is still dear to some scholars in critical race studies and Latino studies who draw theoretical sustenance from the Mexican *indigenista* José Vasconcelos's *Raza cósmica*. More than any other Latin American work on race, *Raza cósmica* trumpets the hybridity model attached to the Latin American racial order and ridicules the purity model attached to the United States.[24] Nevertheless, and despite postmodern attempts to rehabilitate Vasconcelos's modernist racial utopia from the left (Anzaldúa; Martín Alcoff) and from the right (Rodríguez), the Mexican Revolution's best-known educator preached a gospel of *indigenismo* and *mestizaje* that was riven with racisms.[25] Indeed, Vasconcelos was drawn to the works of Grant and Stoddard because he fancied himself a eugenicist.[26] Blacks constitute an inferior, servile race, he claims in *Raza cósmica*, while Asians multiply like rodents. Both of these racist beliefs were, of course, in perfect harmony with the Black Scare and the Yellow Peril sweeping across the Americas and Europe at the time. Available in an inexpensive English-Spanish edition, *Raza cósmica* sparks lively classroom debates about the relations between white supremacy and *indigenismo*, between the life sciences and aesthetics, and about the disparate treatment of indigenous and African peoples throughout Latin American history.

Not once did Grant, Stoddard, García Calderón, or Vasconcelos acknowledge the academic theories about yellow (and/or black) peoples as the first peoples in the Americas—theories that were well publicized in the early twentieth-century United States.[27] Grant and Stoddard's successors would confront those theories head on, for they came to believe that white sovereignty over North and South America, and white supremacy throughout the world, turned upon a primeval whiteness—upon Caucasians who had allegedly designed and ruled over pre-Columbian civilizations.[28] Within white supremacist historicism, the racial hybridity (or racial chaos) attached to colonial and republican Latin America would thereafter take a back seat to white Aztlan.

A keen student of Stoddard's and Grant's teachings, Earnest Sevier Cox (1880–1966) was born in Tennessee but did much of his agitating in Virginia. Along with the internationally known pianist John Powell, he founded the Anglo-Saxon Clubs of America in Richmond, Virginia, in fall 1922. With the moral and intellectual support of Grant and Stoddard, Cox contributed mightily to the passing of the Racial Integrity Law of Virginia and the Johnson-Reed Act that restricted immigration at the federal level, both from 1924.[29] Like Grant and Stoddard in the North, Cox was enamored with modern science. "Beginning in the Progressive Era," Dorr has observed, "eugenics provided generations of educated, self-consciously modern Virginians with a new method of legitimating the South's traditional social order."[30] Cox distributed a copy of his *White America: The American Racial Problem as Seen in a Worldwide Perspective* (1923) to every member of the U.S. Senate, and it would be republished into the 1960s. Two interconnected aspects of *White America* are crucial to white supremacist historicism around pre-Columbian Latin America in the early twentieth century: the drawing of color lines in the Far East, and the notion (entirely absent in Grant and Stoddard) that the founders of pre-Columbian civilizations were Whites/Caucasians /Aryans.[31]

Throughout his influential polemic, Cox flaunts his familiarity with modern academic findings on race in order to vouchsafe his personal delusions of white grandeur. "As a result of recent developments in anthropology and allied sciences," he avers, "we may conclude that all civilizations have arisen from the white race." For Cox, the Aztecs, Incas, and Mayas embodied "the relation of the early white man" to pre-Columbian America. Their civilizations were "Caucasian-like" because the Asians (or "Mongolians") who had gone to the New World in ancient times were not all "yellows."[32] The Aztec, Maya, and Inca cultures had Caucasian origins; they had been designed and built out by whites:

> Those who oppose the theory of the Caucasian origin of the civilizations of Central and South America in reality ask us to believe that the yellow turned megalithic architect in America! That here, but nowhere else in the world, he built mighty pyramids as did the whites of Egypt! That here, but nowhere else, he evolved a religion like the whites of India, the Mediterranean shores, and East Asia! That the creative elements in the pyramid builders of Mexico and Peru were highly specialized Mongolians without the slant eye and short nose, who, in some way not explained, acquired these non-racial characteristics and, in some way equally mysterious, aped the spiritual and material culture of the white race.[33]

Cox states here in a note: "Read the *Secret of the Pacific*, by Reginald Enoch, or the articles on Mexico and Peru in the *Encyclopedia Britannica*. These references will give a summary of the arguments in favor of the theory of Caucasian origin of the cultures of early America."[34] So what does the 1911 edition of the *Encyclopedia Britannica* say about Mexico and Peru? On Peru it says nothing about Caucasians or whites; it says that almost nothing historically is known about the origins of the Incas. On Mexico, the entry writer noted that the "accurate and experienced Alexander von Humboldt considered the native Americans of both continents to be substantially similar in race-characters." "Moreover," the unidentified author wrote, "there are details of Mexican civilization which are most easily accounted for on the supposition that they were borrowed from Asia. They do not seem ancient enough to have to do with a remote Asiatic origin of the nations of America, but rather to be results of comparatively modern intercourse between Asia and America."[35] The upshot is that the "native Americans of both continents" (i.e., from North and South America) are descended from Asian peoples. However, the entries do not affirm that the native peoples of America were Caucasians, as Cox was to claim in *White America*.

Cox's other reference was to British geographer and explorer C. Reginald Enock's *The Secret of the Pacific: A Discussion of the Origin of the Early Civilisations of America, the Toltecs, Aztecs, Mayas, Incas, and Their Predecessors; and of the Possibilities of Asiatic Influence Thereon* (first published in 1912). Enock reviewed the thesis put forward by Charles Darwin's comrade and popularizer Alfred Russell Wallace, who argued that in the southeastern peninsula of Asia there had once existed several Mongol races ruled "by a highly civilised

Figure 6.2 Reginald Enock, *The Secret of the Pacific: A Discussion of the Origin of the Early Civilisations of America, the Toltecs, Aztecs, Mayas, Incas, and Their Predecessors; and of the Possibilities of Asiatic Influence Thereon*
Source: Public domain.

superior race of undoubted Caucasian type." These were the Khmers of Cambodia, according to Wallace, who had mixed with other races to form "the fine Mahori race, whose offshoots reached South America and were the origin of the Incas of Peru and Bolivia." Enock also claimed that Ainos from Japan had sent immigrants to the Americas. Still, his own theory about "the Mongolian connection" contradicted Wallace's and Cox's respective claims about the first peoples in the Americas: "The illustrations given in this book of natives of the Quechua districts of Peru, may, as before mentioned, be compared with the Tibetan and Mongolian faces. Wanderers from Tibet might, indeed, have felt at home in Peru, with its remote towns on lofty plateaus, in the heart of snowy mountains: a land so similar to their own." His photos of yellow or yellow-brown peoples argued for the very Mongolian origins that the myth of a white pre-Columbian America foreclosed.[36]

The fabrication of a primeval whiteness for the Americas was tangled up with the question of color in Asia, ancient and modern. Cox was convinced that "Manchus, Mongols, Koreans and Japanese are admitted to be partly Caucasian, the mongrel descendants of a Neolithic, and possibly also a Paleolithic movement of the Caucasian across northern Asia."[37] That the Chinese, Japanese, Koreans, and other Asians had once been divided into black, white, and yellow segments is a staple of white supremacist discourse in the early twentieth-century United States. It relied on, and to varying degrees manipulated, the discourse on race produced by modern anthropologists, historians, and other social scientists. Three highly regarded authorities in Cox's times were Dudley Buxton, Robert Anderson, and Roland Dixon.

Dixon was a Harvard anthropologist who believed that Stoddard and Grant had mischaracterized the threat posed to whites by the "rising tide of color." Yes, one subtype of the white race—Nordics—was disappearing, but this process began long before the rise of the Japanese and the Chinese: Alpines were gradually overtaking the Nordics just as the Nordics had overtaken the Mediterraneans.[38] Of even greater significance is the fact that Dixon's race model in his academic bestseller *Racial History of Man* (1923) detached phenotype from genotype, so that some "yellows" or "browns" in the Far East were to be categorized as Caucasians:

> In the case of the "yellow" and "light-brown" peoples, who, if our analysis is correct, are, despite their color, derived from the same great fundamental types as the peoples of Europe, the danger is far greater. In them lie latent many of the qualities and abilities which have made European

civilization what it is—not all, however, nor in so full a measure perhaps, yet enough, in the event of their full development, to force upon the peoples now and for so long dominant the most terrible struggle for supremacy they have ever had.[39]

If Dixon was correct—if the populations categorized as yellows (Mongoloids) and browns (Negroids) by Cox were descended from the same stock as Caucasians—then the so-called Yellow Peril represented a different, and far greater, danger than politicians and coffee-table eugenicists were selling to the public. The three racial subtypes of Europe, according to his view, were at risk of losing their "White world-supremacy" to *other whites who did not look like whites*, and Cox and his predecessors were wrong in making "Caucasian" (or "Western European") and "white" synonymous.

Dixon's claims are useful for opening up classroom discussions on the social construction of race, including questions regarding racial nominalism and racial models for the Americas. When the mulatto category was suppressed in the 1930 U.S. Census, did mulattoes cease to exist in the United States?[40] Do different names and categories— different taxonomic practices—in the United States and Latin America constitute different racial realities? Can people who "look" Asian, as Dixon affirmed, actually be white? If race is only a social construction, why do most people (many of them academics) say that Latin America "looks" more racially mixed than the United States, and why do white supremacists fear that the United States is starting to "look" like Latin America?

Another scientific authority in the early twentieth century went even further than Dixon in questioning the whiteness of the first peoples in the Americas. In 1925, the physical anthropologist L. H. Dudley Buxton published *The Peoples of Asia*, which was to become a standard reference work on Asia for several decades. In discussing the different races of Asia, Buxton wrote that there were basically three: blacks, yellows, and whites. He noted great variation within each race, especially in the yellows (or Mongolians):

> The skin-colour varies considerably. It practically always has some shade of yellow which may be almost white in some of the high-born ladies of Northern China. The fairest men are usually of a very pale saffron, only a little more yellow than the tint which would be described as olive and often lighter in tone than olive. This tint may perhaps represent the true colouring of the "Sons of Han," but in a large collection of male Chinese it comes not very frequently. The complexion then varies through the

various shades of yellow to a dusky yellow-brown, sometimes so dark that the yellow tinge is almost obscure, and among some of the tropical peoples the brown has almost a greenish hue.[41]

Here Buxton indicates that color variation among the Chinese is tied to environmental factors—most notably, social class and the hours of sun exposure associated with it, and climatic differences between the North and South. Buxton was well aware of Robert Anderson's *Story of Extinct Civilizations of the East* in which he claimed that the color of the first modern humans was most likely yellow because the human race had originated in Asia: "The yellow men have immediately occupied the great central and northeastern plains of Eurasia, and are therefore called Mongols or Turan-Chinese. [...] From their number, position, and other considerations they appear to have first existed; the other two races being derived from them by emigration, change of climate and mode of living." The "ruddy or olive-white Caucasian" and "the brown-black negro" came into being after splitting from the primeval yellows.[42]

Attached to his trans-American myth of primeval whiteness, Cox would not entertain scientific findings that supported radically different migration scenarios. He flattened existing race models whenever it suited him by misrepresenting their complexities and/or willfully ignoring the true intentions of their creators. Both of these tacks come into play as he invokes the authority of Aleš Hrdlička, then curator of the Division of Physical Anthropology at the U.S. National Museum (now the Smithsonian) and the foremost expert on early man in the Americas. "That there is a non-Mongoloid element in the American Indian is apparent," Cox states matter of factly in *White America*. "So Caucasian-like are, or were, some of the tribes, that Alfred Wallace suggested that they came from Europe [...], while anthropologists of high standing, like the ever critical Dr. Aleš Hrdlička, is of the opinion that the elements which go to make up the American Indian, in part, may be traced through Asia into Europe."[43]

In "Remains in Eastern Asia of the Race That Peopled America," Hrdlička explained: "There exist to-day over large parts of eastern Siberia, and in Mongolia, Tibet, and other regions of that part of the world, numerous remains, which now form constituent parts of more modern tribes or nations, of a more ancient population (related in origin perhaps with the latest Paleolithic European), which was physically identical with and in all probability gave rise to the American Indian."[44] However, Cox misrepresented Hrdlička's work, according to which the

Figure 6.3 Aleš Hrdlička, "Remains in Eastern Asia of the Race That Peopled America (with Three Plates)." *Smithsonian Miscellaneous Collections* (1912)
Source: Public domain.

Figure 6.4 Aleš Hrdlička, "Remains in Eastern Asia of the Race That Peopled America (with Three Plates)." *Smithsonian Miscellaneous Collections* (1912)
Source: Public domain.

Figure 6.5 Aleš Hrdlička, "Remains in Eastern Asia of the Race That Peopled America (with Three Plates)." *Smithsonian Miscellaneous Collections* (1912)
Source: Public domain.

first peoples in the Americas appear to have been—and still are in the early twentieth century—yellows and browns: "the American native did not originate in America, but is the result of a comparatively recent, post-glacial immigration into this country; that he is physically and otherwise most closely related to the yellow-brown peoples of eastern Asia and Polynesia; and that in all probability he represents [...] a gradual overflow from north-eastern Siberia."[45] Three plates were meant to illustrate that descendants of the peoples from whom the pre-Columbian civilizations sprang could still be found in parts of Siberia and Mongolia. "A Family of Yenisei Ostiaks" is the caption of the third photo, which Hrdlička associated with the peoples found in the first and second photos. Cox manipulated the anthropologist Hrdlička's findings just as he manipulated the geographer Enock's findings: Old World yellow or yellow-brown peoples lost their melanin in *White America*.[46]

Definitional challenges arise in classroom discussions of *White America*. If color was the folk concept of race, was whiteness biological? What, exactly, was whiteness to Cox and others? On the one hand, whiteness was not a biological fact: It was a cultural construct like Western European or Caucasian or Aryan. On the other hand, it *was* a biological category in that white supremacists insisted on the whiteness (not the yellowness or brownness) of Western European culture. This seems rather banal today because we have forgotten that in the first half of the twentieth century white was not the only option for the originators of Western European culture: Dixon was one of the foremost authorities on the history of race well into the 1940s, and Anderson, too, was required reading for students of the Asian world. By giving environment and heredity equal shares in the forging of human varieties known as races, their race models undermined white supremacy's color-coding by severing the relationship between genotype and phenotype.

It is worth emphasizing to students that the dark slave–pale master paradigm so very central to white supremacist discourse in the early twentieth century was flexible enough to assimilate Asian slaves of indeterminate hues into the narrative of the white man's westward trek across history: "The way before the Caucasian was easy and enticing," Cox reasoned. "It was peopled by inferiors that the Caucasian's race-long history had taught him he could subdue and enslave. Did the Caucasian not follow the colored to America as he did to central Africa and southern Asia [...]?"[47] America was more easily subjugated than Europe, Cox asserted, because in Europe "there were not servile hordes of Negroid and Mongoloid peoples among whom the Caucasians could reign as aristocrats and thus gain opportunity for intellectual pursuits."[48] Again, it is clear that the coexistence and/or codependence of the Black Scare (negrophobia) and the Yellow Peril deserve far more attention than scholars in critical race studies have paid to them. Hatred for "yellows" and "yellow-browns" rivets Cox's study no less than his hatred for blacks and mulattoes.

Students do not fail to notice how Cox reiterates that the servile majority of the first Asian peoples in the Americas were "colored"— Mongoloids ("yellows") and Negroids ("browns")—while the commanding minority were "early Caucasians," or whites. It is productive to ask students to explain how this oft-repeated claim sets up Cox's larger argument that pre-Columbian civilizations collapsed due to miscegenation between white ruler and "colored" subject, and why the tale of an ancient white America lost to miscegenation would be welcomed by opponents of immigration and integration in the Jim Crow era.[49] This second question serves as a transition to so-called "theories" about white Aztlan in the digital age.

In the late twentieth century, and the early twenty-first, white supremacy has evolved by recycling the ideas of their forebears and redeploying them in new racial projects. Today's white supremacists prefer to use the Internet rather than books and pamphlets, but they mimic Stoddard, Grant, and Cox's manipulations of academic research on human variation. Links to articles on race and genetics published in

The American Journal of Human Genetics, *Annals of Human Genetics*, *Human Immunology*, and *Human Biology* appear frequently on white supremacist websites.[50] Still another ideological stratagem that unites postmodern and modern white supremacists is their devotion to making Spaniards white. For Grant, Stoddard, and Cox, Spaniards were Mediterraneans who took their place alongside the other white peoples of Europe: the Alpines and the Nordics. For postmodern white supremacists, the Spanish and the Portuguese are once-were-White epitomes of miscegenation and its evils: "lessons in decline."[51] On the Stormfront White Nationalist Community website, the overlay of their domain name (stormfront.org) on a white marble bust of Isabella of Castile is a visual insinuation of America's primeval whiteness: Christianity, maternity, and racial integrity become fused in America's beginnings through this Spanish queen. She and her husband Ferdinand—more commonly known as "Catholic Monarchs" (*Reyes Católicos*)—supported Columbus's planned voyages to Asia that would take him instead to the New World, which geneticists now believe was first settled by Asians.

Digital white supremacists share their predecessors' goal to reinforce the Color Line by restricting immigration and discouraging integration in the United States. Unlike Grant, Stoddard, and Cox, however, they target primarily Latin Americans and Latinos. On www.puttroopsontheborder.com, a website affiliated with the white supremacist theology movement known as Christian Identity, there is a link titled "Alien Invasion" that transfers Grant's fear of Southern European, Jewish, and West Indian immigrants to Latin Americans and Latinos.[52] On another link, Aztlan is reclaimed for whites under the heading that asks rhetorically, "Is their claim to America Legitimate?" In a brief section titled "The Basis of Their Right to Reclaim?," the unnamed author ridicules Chicano activists (especially MeCha, the foremost organization of Chicanos on the left) and provides a link to "Plan de Aztlan," a manifesto to repossess Aztlan, which many Chicanos revere as their ancestral homeland. Still another link takes the user to "Another Theory on Aztlan," rooted in the racial theology of Christian Identity:

> Despite academia's modern claim that the white race are the collective descendents [sic] of European barbarians and Cro-Magnum cave dwellers, nothing can be more further [sic] from the truth. [...] From the cradle of civilization in Mesopotamia, to Egypt, the Orient, the Pacific Isles, and India the creative and stabilizing force of the white race can be found in the ancient writings, scientific discoveries, relics, and ruins.

> The same can be said for North and South America. But as a non-white people are now claiming to have originated the once great South American empires and the pyramids, temples, and highways of Mexico and thus believe the white man must turn it over to them, we will briefly digress to show the absurdity that these people created any great empire, but are rather the result of another integration nightmare and have no sacred right to land that the brave men at the Alamo fought and died for![53]

Heaping scorn on academia and reaffirming what Stoddard called *White World-Supremacy*, Christian Identity's so-called "theory" echoes Cox's claims about pre-Columbian America and silences recent genetic evidence in support of the Asian settlement theory. Moreover, the phrase "another integration nightmare" manifests the rabid negrophobia of postmodern white supremacists who view their forerunners' fears of miscegenation as both prescient and unacknowledged by academia, the government, and the white public at large.

The unnamed author goes on to argue that nonwhites lived in the Americas before whites did, but the founders of pre-Columbian civilizations were white men. Those ancient white mariners came, saw, conquered, and then simply vanished, not through miscegenation (as Cox had argued), but through exile or abandonment:

> It is quite accurate for the non-white people of Mexico and South America to make the claim that they were here before the white man—after all, the white man and woman (Adam and Eve) came from the Tarim Basin and Mesopotamia region. However, it is equally inaccurate for them to make any claim to our Southwestern states because it was once known as Aztlan. Aztlan it once was—but it was not of their creation. After the white man and woman left, the great cities, highways, temples, market places, scientific centers, and essentially all civilization eroded until the white Spaniards came to find a destitute, bloodthirsty, cannibalistic people who had not advanced one bit since the founders of their cities had left them. The battle of the Alamo settled the reconquering of the South West or Aztlan. It was reconquered by the people who had made it great once before—the white race![54]

In studying and teaching white supremacist interpretations of pre-Columbian history such as the myth of white Aztlan, it quickly becomes clear that white supremacists try to shut down any and all discussion of competing Latin American and U.S. narratives of the Mexican-American War. The attempt registered earlier affords students an opportunity to discuss the Mexican-American War and the American Civil War in ways that draw the cultures of Mexico closer to Native Americans, African

Americans, and Latino Americans in the United States. Selections from Jack Forbes's study, and more recently, Rachel Adams's essay on *Lone Star*, along with the movie itself, can ground discussion about the overlappings of Native American, African American, and Latino history in the Southwestern United States, and about slavery's role within the Mexican-American War and the American Civil War.[55]

Another engaging reading, this one by a Latin American, is the Argentinean President Domingo F. Sarmiento's little-known speech to the Historical Society of Rhode Island, "North and South America" (1865). In several passages from this essay, written when he was Argentina's representative in the United States (1865–1868), Sarmiento ties the downfall of the United States to imperialism as manifested in the Mexican-American War and, ultimately, to slavery, which he blamed for the Mexican-American War and the American Civil War. If students can read Spanish, Sarmiento's introduction to his biography of President Lincoln contains several pages on the same topic that make clear just how much was left unsettled by the white man at the Battle of the Alamo, much to the consternation of Christian Identity and other white supremacists today. Excerpts from Brack's and Robinson's respective accounts are also useful background reading for class discussions on this topic.[56]

Another link on www.puttroopsontheborder.com leads to "A Biblical and Historical Look at the Immigration Crisis," written by a Mrs. Rachel Pendergraft, in which South America is listed as the most tragic example of great ancient civilizations that collapsed due to miscegenation. Here again it is alleged that the white man founded the Aztec and Maya civilizations, which would thereafter decline into savagery: "The Aryan/Caucasian influence on South America can be seen in the remains of the ancient pyramids, ancient highways, and temples. The race mixing as in Egypt and India thousands of years ago toppled a once advanced civilization. When the white Castilian Spanish came to South America in the 1500s they found a barbaric culture of human sacrifice and vile superstition amongst ancient ruins."[57] Two of Christian Identity's privileged authorities on pre-Columbian history are Madison Grant and Earnest Sevier Cox, and Mrs. Pendergraft relies on both in her article.

In my teaching and research on race in the Americas I strive to eschew theoretical paradigms and assumptions that threaten to simplify rather than clarify how the floating signifiers of race and class differ from and resemble each other in different periods and places. Although it is catchy to claim, as Latino cooks in Emilio Estévez's *Bobby* (2006) do, that Latinos "are the new niggers," it is impossible to delineate where the Black Scare (or negrophobia) ends and the Brown

Scare (or Latinophobia) begins for twenty-first-century white supremacists who subscribe to the racial theology of Christian Identity.[58] In this regard, they ape early twentieth-century white supremacists: Grant's morbid fear of West Indian immigration, alongside his joy at the return of Mexican (Indian) migrant workers to their birth country, proves that negrophobia and Latinophobia co-existed in the early twentieth century just as virulently as they co-exist in the "another integration nightmare" scenario of cyberspace white supremacists. Furthermore, the possibility that Latinos, like European immigrants, will become "honorary Whites," has also become part of the public discourse on Latinos, which no doubt terrifies white supremacists (and many Latinos) today.

And yet, this very possibility calls to mind not only studies by Jacobson and Roediger, but also James Baldwin's *The Evidence of Things Not Seen*, which (along with Toni Morrison's *Playing in the Dark*) poured the foundations for critical whiteness studies, a growing subfield of critical race studies. Ostensibly about the Atlanta child murders and the Wayne Williams trial, Baldwin's essay picks up where *The Fire Next Time* left off: It is an unsparing meditation on white supremacy's structuring of the history of the West.[59] His intellectual and moral clarity regarding the intersections of class and race in the United States foreshadow the main lines of Roedigers's argument in *Wages of Whiteness* and bring into relief how white supremacy structures the American Dream of immigrants:

> This species of folklore—out of which Horatio Alger, among others, was to make a killing—does not imply, but very clearly states that America is the land of opportunity and that Blacks, therefore, deserve their situation here. I no longer really care why the authors of this self-serving fantasy cling to it so ignobly. It is, nevertheless, worth pointing out that this fable tells us, simply, that the economic and political base of the city was determined entirely by immigrants who were in the process of becoming White Americans. Whatever differences the Irish, Greek, Pole, Italian, Finn, Norwegian, German—as well as Jews, from all over the world—may have had between themselves, they never, as entities, differed among themselves concerning the role and utility of the Black. They could not afford to: those who dared were hounded out of the White community (for it became *White* whenever *Black* was mentioned) as being *worse* than niggers—as being traitors, that is, to the American Dream. [...][60]
>
> That community that was in the process of becoming White could—and did—always bury its differences long enough to make certain that the Black could not rise to a place of sufficient recognition to threaten the structure of the labor union or the city or the state. And the saddest thing about this is that, even by the time I came along, searching for a

watermelon in the streets of Harlem, there was nothing wicked about the White people who still lived in Harlem. But they could not see to what extent they themselves—having been manipulated into becoming White—were being manipulated by interests that cared no more about their lives than they cared about the lives of niggers.[61]

Baldwin's analytical perspicacity brings to light that whiteness and blackness were New World inventions inextricably linked to voluntary immigration and involuntary immigration (indentured servitude, chattel slavery), themselves the pivots of a dialectics of European capitalism and white supremacy. Further, his emphasis on the intersections of race and class in the production of whiteness in the United States allows students and scholars alike to perceive a signal difference between the well-educated and well-heeled Grant and Stoddard and their postmodern brood: The majority of twenty-first-century white supremacists are not aristocrats or the scions of political and cultural elites. Rather, they are working-class white males who feel disenfranchised but are unable to perceive that they are—to slightly alter Baldwin's words—"being manipulated by interests that care no more about their lives than they care about the lives of" disenfranchised African Americans and Latinos. It is precisely this group of whites to whom cyberspace white supremacists pitch their products—articles, video games, music, speeches, and so on.

White supremacists, Christian extremists, antigovernment groups, racialists, white nationalists—all of these terms raise definitional conundrums elided in this essay.[62] Why are Christian Identity adherents characterized in public discourse as "antigovernment" groups and not as "Christian extremists" or "Christianists" or even "Christian terrorists"? They are clearly a different kettle of fish from nontheological white supremacists past and present. Their nonelite educational and economic backgrounds also distinguish them from racists such as Samuel Francis, Patrick Buchanan, and Jared Taylor who self-identify as, or affiliate themselves with, "racialists," "white nationalists," "Euro-American patriots," or "white-consciousness-raisers." Indeed, the differences in education and occupation—in social class—between the latter and self-avowed white supremacists such as Aryan Nations or Stormfront parallel the differences (and denials) between members of the early twentieth-century Anglo-Saxon clubs and the Ku Klux Klan.[63] Moreover, white supremacists in the early twentieth-century United States and Western Europe considered themselves culturally and economically superior to the masses, just as Buchanan and his colleagues consider themselves superior to hate groups like Aryan Nations or Stormfront.

In this essay I have proposed that white supremacism is not, and has never been, homogeneous—as monochrome or as monotone—as its proponents are wont to claim. While recognizing their material differences, I have pointed out numerous continuities between the white supremacists of the early twentieth century and those of the present. Among these affinities are their shared desire to appear (post)modern and scientific: The early twentieth-century cadre did so by engaging with eugenics and the social sciences in order to flatten the complexities of race models and invent a distinct color (white) and a distinct culture (Aryan or Caucasian or European) that often were incongruent with those same models; the second generation also distorts findings in the natural and social sciences and at once resorts to digital technology's capacity to serialize and remediate. Both groups have represented the Spanish and the Portuguese as pioneers of "White Pride Worldwide" and at once "lessons in (racial) decline." They have also shared a need to color-code the prehistory of the Americas, and put forth a pessimistic soothsaying about Latin America and the integration of immigrants in the United States perceived as members of inferior, nonwhite races.

More broadly, my approach to teaching and studying white supremacist historicism around the Americas challenges three patterns within the burgeoning field of critical race studies: (1), the critical posture that severs Asian American history from African American history, the Yellow Peril from the Black Scare; (2), the assumption that Latinophobia and negrophobia have led entirely separate lives in the modern and postmodern United States; and (3), the racial purity versus racial hybridity paradigm that both reflects and inflects racial formations in the United States and Latin America whenever the Americas are studied comparatively. Racial ordering occurs within scholarship as well as outside of it, and to embrace that hemispheric contrast or third pattern—to invest racial nominalism with immanence—is not only to impose order on our objects of study, but also (and far more disturbingly) to impose an order that grants white supremacists supreme powers: in effect, to concede that the mixed America is "over there" and the racially intact (i.e., white) America is (and always has been) here. In the wake of the historic 2008 U.S. presidential election, which has reinvigorated white supremacist organizations throughout Europe and the United States, it is especially important that we and our students possess intellectual and moral clarity in our paradigms.

Notes

1. Rodriguez's *Brown* is a conservative's foray into the racial history of the Americas that adheres closely to the contrastive model. On the left,

Martín Alcoff's commitment to the same approach is evident in several studies: "[I]n Latin America mixed-race persons do not create a cognitive crisis because they are the norm. There, racial identity is determined along a continuum of colour without sharp borders;" "My original entry into this area of work was motivated by a concern to understand and in some sense validate hybrid identity or hybrid positionality against purist, essentialist accounts. And the motivation for this was the felt alienation of having a mestizo identity (normative in Latin America and the Caribbean) but living in a purist culture (the USA) [...]." See Richard Rodriguez, *Brown: The Last Discovery of America* (New York: Viking, 2002); Linda Martín Alcoff, "Towards a Phenomenology of Racial Embodiment," *Radical Philosophy* 95 (May–June 1999): 25–26, n. 11 (first quotation) and Linda Martín Alcoff, "Philosophy and Racial Identity," *Radical* Philosophy 75 (1996): 9 (second quotation).
2. See Ann Burlein, *Lift High the Cross: Where White Supremacy and the Christian Right Converge* (Durham: Duke University Press, 2002).
3. Indeed, according to the current generation of CRT proponents, the roots of CRT are to be found in anthropology, history, gender studies, sociology, and literary studies. See Charles R. Lawrence III, "Foreword" in *Crossroads, Directions, and a New Critical Race Theory*, eds. Francisco Valdes, Jerome McCristal Culp, and Angela P. Harris (Philadelphia: Temple University Press, 2002).
4. On the French, U.S., and Latin American contributors to this paradigm for understanding race in the nineteenth-century Americas, see Ruth Hill, "Entre lo transatlantic y lo hemisférico: Los proyectos raciales de Andrés Bello," in Eyda Merediz and Nina Gerassi Navarro, eds., *Más allá de lo transatlantic*, special issue of *Revista Iberoamericana* 75 (Julio–Septiembre 2009): 719–735.
5. A sturdy outline of the variegated ideological strands of *mestizaje* is found in Peter Wade, "Rethinking Mestizaje: Ideology and Lived Experience," *Journal of Latin American Studies* (2005): 37.2: 239–57. What Outlaw has affirmed a propos the melting pot myth is equally true of *mestizaje* when "whites" is substituted for "Anglo-Saxons": "Though this notion is to some degree progressive (for example, it promotes the mediation of differences that might block the achievement of a unified society of diverse peoples), it has tended to mask the realities of ethnic, racial, and national pluralism and the hegemony of Anglo-Saxons." See Lucius T. Outlaw, Jr., "Philosophy, African-Americans, and the Unfinished American Revolution," in *On Race and Philosophy* (New York: Routledge, 1996): 47.
6. Daniel Pick, *Faces of Degeneration: A European Disorder, c. 1848–1918* (Cambridge: Cambridge University Press, 1993); Matthew Frye Jacobson, *Whiteness of a Different Color: European Immigrants and the Alchemy of Race* (Cambridge, Mass.: Harvard University Press, 1998).
7. See Francisco Bilbao, *El Evangelio Americano. Obras Completas*, Vol. 2, ed. Manuel Bilbao (2 vols., Buenos Aires: Imprenta de Buenos Aires, 1865);

Victor Arreguine, *En qué consiste la superioridad de los latinos sobre los anglosajones* (Buenos Aires: Enseñanza Pública, 1900).

8. See Matthew Pratt Guterl, *The Color of Race in America, 1900–1940* (Cambridge, Mass.: Harvard University Press, 2001); Jacobson, *Whiteness of a Different Color*, and Claudette Bennett, "Racial Categories Used in Decennial Censuses, 1790 to the Present," *Government Information Quarterly* 17 (2000): 161–180.

9. Madison Grant, *The Passing of the Great Race, or The Racial Basis of European History* (New York: Charles Scribner's Sons, 1916); Madison Grant, *The Conquest of a Continent or The Expansion of Races in America* (New York: Charles Scribner's Sons, 1933); Lothrop Stoddard, *Rising Tide of Color Against White World-Supremacy* (New York: Charles Scribner's Sons, 1921); Lothrop Stoddard, *Clashing Tides of Colour* (New York: Charles Scribner's Sons, 1935). Also, Michel Foucault and, more recently, Linda Martín Alcoff, "Philosophy," have underscored the relationship between racial taxonomies and mapping in early modern Europe, while Guterl has made clear how the two are related in the early twentieth-century United States. See Michel Foucalt, *The Archaeology of Knowledge*, trans. A. M. Sheridan Smith (New York: Pantheon, 1982); Alcoff, "Philosophy and Racial Identity."

10. See Jennifer Michael Hecht, "Vacher de Lapouge and the Rise of Nazi Science," *Journal of the History of Ideas* 61 (April 2000): 285–304; Johnathan Peter Sapiro, *Defending the Master Race: Conservation, Eugenics, and the Legacy of Madison Grant* (Lebanon, N. H.: University Press of New England, 2009).

11. Lothrop Stoddard, *Into the Darkness: Nazi Germany Today* (New York: Duell, Sloan, & Pearce, 1940), chapter 17.

12. See Roland B. Dixon, *The Racial History of Man* (New York: Charles Scribner's Sons, 1923), 520–521; L. C. Dunn and Theodosius Dobzhansky, *Heredity, Race, and Society* (New York: Penguin, 1946), 8; and Clyde Kluckhohn, *Mirror for Man: A Survey of Human Behavior and Social Attitudes* (New York: Premier Books, 1957), 107.

13. Grant, *Passing of the Great Race*, 14–15. As Jacobson's and Reginald Horsman's respective studies have shown, the capacity for self-government was frequently correlated to racial or ethnic superiority (specifically, Anglo-Saxonness) in the nineteenth century. See Jacobson, *Whiteness of a Different Color*; Reginald Horsman, *Race and Manifest Destiny: The Origins of American Racial Anglo-Saxonism* (Cambridge, Mass.: Harvard University Press, 1981).

14. Grant, *Conquest of a Continent*, 223–234, 346. Grant writes: "While the census of 1930 counted nearly a million and a half Mexicans in the United States, it is probable that the number has since then diminished, and it is of highest importance that it should not be allowed to increase. The Mexican Indian has no racial qualities to contribute to the United States population that are now needed, and if he has any cultural contribution to

make it will not be made by the immigration of hundreds of thousands of illiterate and destitute laborers"(330).
15. Stoddard, *Clashing Tides of Colour*, 110 (for a discussion of oligarchy), 114 (for Aryanization), and 116 (block quotation).
16. On Stoddard and the Yellow Peril more generally, including his influence on the late Samuel Huntington's "clash of civilizations" thesis, see Robert G. Lee, "Brown is the New Yellow: The Yellow Peril in an Age of Terror," in James T. Campbell, Matthew Pratt Guterl, and Robert G. Lee, eds., *Race, Nation, & Empire in American History* (Chapel Hill: University of North Carolina Press, 2007): 335–351.
17. Stoddard, *Rising Tide of Color Against White World-Supremacy*, 130.
18. Frank H. Wu, *Yellow: Race in America Beyond Black and White* (New York: Basic Books, 2003).
19. Stoddard, *Rising Tide of Color Against White World-Supremacy*, 229.
20. Stoddard, *Rising Tide of Color Against White World-Supremacy*, 232.
21. Francisco García Calderón, *Les Democraties Latines* (Paris: Flammarian, 1912).
22. Edward A. Bradford, "What Latin Americans Think of Themselves and Us: F. García Calderón, Peruvian Diplomat, Gives a Clear View of the Ambitions and Anxieties of Our Neighbors to the South in a Notable Book, Which Has a Preface by President Poincaré of France," *New York Times* Online, February 23, 1913 (accessed December 17, 2008).
23. Bradford, "What Latin Americans Think of Themselves and Us."
24. José Vasconcelos, *The Cosmic Race/La raza cosmic,* trans. Didier T. Jaén (Baltimore: Johns Hopkins University Press, 1997).
25. See Gloria Anzaldúa, *Borderlands/La Frontera: The New* Mestiza (San Francisco: Spinsters/Aunt Lute Book Company, 1987); Martín Alcoff, "Philosophy and Racial Identity;" Martín Alcoff, "Towards a Phenomenology of Racial Embodiment;" Rodriguez, *Brown*.
26. Vasconcelos was familiar with works by Grant, Stoddard, and other U.S. white supremacists. In *Indología* (first published in 1926), he addresses Grant's *Passing of the Great Race* in a footnote spanning seven pages. Vasconcelos applauds Grant's views on population control, using figurative language that recalls Stoddard's *Rising Tide of Color*: "Before concluding these statements, I should affirm that I am totally in agreement with Grant when he condemns the marriages between inferior and degenerate types. I have fervently preached the gospel of birth control to the masses of my country. The oppressed races endure fertility's lash, but this curse turns against their oppressors and ends up drowning them" (my translation). See José Vasconcelos, *Indología: Una interpretación de la cultura ibero-americana* (Paris: Agencia Mundial de Librería, 1928), 102–108); Grant, *The Passing of the Great Race, or The Racial Basis of European History*, 107.
27. See Aleš Hrdlička, "Remains in Eastern Asia of the Race that Peopled America (with Three Plates), *Smithsonian Miscellaneous Collections* 60

(1912): 1–5; Roland B. Dixon, "Says Negroid Group Discovered America: Dr. Dixon Startles Scientists Asserting Black Types Were Among Indian's Ancestors," *New York Times* Online, December 30, 1922 (accessed March 4, 2009); Dixon, *Racial History*; and Leo Wiener, *Africa and the Discovery of America* (3 vols., Philadelphia: Innes & Sons, 1920–1922).
28. As Sayre observes a propos the Kennewick Man controversy: "Euro-Americans who call themselves 'Caucasian' [...] are eager to assert kinship [...] because such an affiliation supports myths of Europan sovereignty in North America, and marginalizes the sovereignty of American Indians." Gordon M. Sayre, "Prehistoric Diasporas: Colonial Theories of the Origins of Native American Peoples," in Phillip Beidler and Gary Taylor eds., *Writing Race Across the Atlantic World: Medieval to Modern* (New York: Palgrave Macmillan, 2005), 53.
29. See Derryn E. Moten, "Racial Integrity or 'Race Suicide': Virginia's Eugenic Movement, W. E. B. Du Bois, and the Work of Walter A. Plecker," *Negro History Bulletin* 62 (April–September 1999): 6–17; Gregory Michael Dorr, "Assuring America's Place in the Sun: Ivey Foreman Lewis and the Teaching of Eugenics at the University of Virginia, 1915–1953," *Journal of Southern History* 66 (February–March 2000), 257–296; J. Douglas Smith, "The Campaign of Racial Purity and the Erosion of Paternalism in Virginia, 1922–1930: 'Nominally White, Biologically Mixed, and Legally Negro,'" *Journal of Southern History* 68 (February 2002): 65–106.
30. Dorr, "Assuring America's Place in the Sun," 259.
31. Earnest Sevier Cox, *White America: The American Racial Problem as Seen in a Worldwide Perspective* (Richmond: White America Society, 1920).
32. Cox, *White America*, 154, 160.
33. Cox, *White America*, 161.
34. Cox, *White America*, 85, n. 3.
35. "Peru," *Encyclopedia Britannica*, 11th ed. (1911), 274; "Mexico," *Encyclopedia Britannica*, 11th ed. (1911), 330.
36. C. Reginald Enock, *The Secret of the Pacific: A Discussion of the Origin of the Early Civilisations of America, the Toltecs, Aztecx, Mayas, Incas, and Their Predecessors; and of the Possibilities of Asiatic Influence Thereon* (New York: Charles Scribner's Sons: 1912), 300–301 (first and second quotation), 233 (third quotation), 303–304 (fourth quotation).
37. Cox, *White America*, 156.
38. Dixon, *Racial History of Man*, 520.
39. Dixon, *Racial History of Man*, 520.
40. Bennett, "Racial Categories Used in Decennial Censuses, 1790 to the Present."
41. L. H. Dudley Buxton, *The Peoples of Asia* (New York: Alfred A. Knopf, 1925), 60–61.
42. Robert E. Anderson, *The Story of Extinct Civilizations of the East* (New York: The University Society, 1896), 14–15.
43. Cox, *White America*, 163–164.

44. Hrdlička, "Remains in Eastern Asia of the Race That Peopled America," 4.
45. Hrdlička, "Remains in Eastern Asia of the Race That Peopled America," 1.
46. Poole's account is rich in suggestions for a separate study of these photographs as (neo)colonial artifacts, but my interest here lies in how Cox deliberately misreads (and "misviews") the scientific literature on race that discredits his white supremacist confections. In this regard, it is telling (perhaps even confessional) that Cox chose not to reproduce any of the photos that he misrepresents as photos of Caucasians/whites/Aryans.
47. Cox, *White America*, 160.
48. Cox, *White America*, 49.
49. Cox, *White America*, 161.
50. See the scientific literature referenced in Stormfront's "Racial Ancestry in Iberia (Spain)," http://www.stormfront.org, Stormfront White Nationalist Community, 1–9.
51. See Stormfront's *March of the Titans: A History of the White Race*, ch. 22, "Lessons in Decline Part One: Spain," 1–12; "Lessons in Decline Part Two: Portugal," 1–8.
52. "Alien Invasion" http://puttroopsontheborder.com. On the Stormfront website there appears a link to www.puttroopsontheborder.com. Throughout the writings published on this site there abound references to Christianity and to Christian Identity's teachings, though Don Black (the publisher of Stormfront) is not usually categorized as a Christian extremist. Walters notes that the racial theology elaborated on by Christian Identity (or, simply, Identity) has followers among groups as diverse as Aryan Nations, the Ku Klux Klan, and the Montana Freemen, a point reinforced by Schlatter. See Jerome Walters, *One Aryan Nation Under God: Exposing the New Racial Extremists* (Cleveland: Pilgrim Press, 2000), 1–7; Evelyn A. Schlatter, *Aryan Cowboys: White Supremacists and the Search for a New Frontier 1970–2000* (Austin: University of Texas Press, 2006). The best general introduction to Christian Identity remains Barkun, but see also Quarles and, more recently, Burlein. Barkun dates the Identity movement to the 1980s, while Swain and Nieli date white nationalism to the early 1990s. White supremacists who self-identify as Christians often group themselves regionally, as is the case with the aforementioned Montana Freemen, and with Christian Resistance (whose motto is "committed to the Survival of the Southern People") and Little Geneva Reformed Confederate Theocrats (whose motto is "Creating an Old Kind of Christian"). Michael Barkun, *Religion and the Racist Right: The Origins of the Christian Identity Movement*, rev. ed. (Chapel Hill: University of North Carolina Press, 1997); Chester L. Quarles, *Christian Identity: The Aryan American Bloodline Religion* (Jefferson, N. C.: McFarland and Company, 2004); Burlein, *Lift High the Cross*.
53. "Another Theory on Aztlan," http://puttroopsontheborder.com, 3–4.
54. "Another Theory on Aztlan," 8–9.

55. Jack Forbes, *Black Africans and Native Americans: Color, Race, and Caste in the Evolution of Red-Black Peoples* (Oxford: Oxford University Press, 1988); Rachel Adams, Blackness Goes South: Race and *Mestizaje* in Our America," in Sandhya Rahendra Shukla and Heidi Tinsman, eds., *Imagining Our Americas: Toward a Transnational Frame* (Durham: Duke University Press, 2007).
56. Domingo F. Sarmiento, "North and South America: A Discourse Delivered Before the Rhode-Island Historical Society, December 27, 1865," (Providence, R. I.:Knowles, Anthony & Co., 1866); Domingo F. Sarmiento, *Vida de Abran Lincoln. Décimosesto Presidente de los Estados Unidos* (New York: A. Appleton and Co., 1866); Gene M. Brack, *Mexico Views Manifest Destiny, 1821–1846: An Essay on the Origins of the Mexican War* (Albuquerque: University of New Mexico Press, 1975); Cecil Robinson, ed., *The View from Chapultepec: Mexican Writers on the Mexican-American War*, trans. Cecil Robinson (Tucson: University of Arizona Press, 1989).
57. Rachel Pendergraft, "A Biblical and Historical View of Our Current Immigration Crisis," http://puttroopsontheborder.com, 9–10.
58. *Bobby*, written and directed by Emilio Estévez (The Weinstien Company, 2006).
59. James Baldwin, *The Evidence of Things Not Seen* (New York: Holt, Rinehart, & Winston, 1985). See also Toni Morrison, *Playing in the Dark: Whiteness and the Literary Imagination* (New York: Vintage, 1993); Baldwin, *The Fire Next Time* (New York: Dell, 1963); Jacobson, *Whiteness of a Different Color*; David R. Roediger, *The Wages of Whiteness: Race and the Making of the American Working Class* (London: Verso, 1991).
60. James Baldwin, *The Evidence of Things Not Seen* (New York: Henry Holt, 1985), 32.
61. Baldwin, *The Evidence of Things Not Seen*, 32–33.
62. Swain and Nieli, for example, use the terms *white nationalist*, *white supremacist*, *racialist*, *white protest group*, *white racial activist*, and *white separatist* indiscriminately, and their collection of interviews includes Stormfront's Don Black, former Klansman David Duke, and the author Jared Taylor, a disciple of the late Samuel Francis (Patrick Buchanan's friend and interlocutor). See Carol M. Swain and Russ Nieli, eds., *Contemporary Voices of White Nationalism in America* (Cambridge: Cambridge University Press, 2003).
63. "While no evidence exists," J. Smith concedes, "to suggest that Powell himself ever belonged to the Klan, his insistence that he and the Anglo-Saxon Clubs were in no way connected with the Klan was simply not true" (4). Several members of the Clubs were former Klansmen, he notes: "By putting a new face on the Klan, Powell was able to legitimize the Anglo-Saxon Clubs in the minds of respectable, elite Virginians" (4). It is highly likely, too, that Klansmen saw the Anglo-Saxon Club as a vehicle for social mobility, although Smith does not suggest this. Smith, "The Campaign of Racial Purity and the Erosion of Paternalism in Virginia, 1922–1930.

CHAPTER 7

The Making of "Americans": Old Boundaries, New Realities

Karen Manges Douglas and Rogelio Saenz

This essay explores the highly complex issues of race and ethnicity in the early-twenty-first-century United States, the context into which millions of people from the other American nations enter. Though the election of the first black person to the U.S. presidency reflects real progress, the racial discourse that both surrounded Obama's election and continues to persist during his administration points to the tremendous work on race that still needs to be done. We illustrate the significant role race has played historically in influencing the laws meant to control the population coming to the United States, contorting the distribution of property and resources. We then examine how these race-conscious policies continue to play out into the twenty-first century.

While U.S. race relations remain driven by a framework revolving around two races—black and white—the demographic reality challenges this model. Bonilla-Silva and his colleagues (Bonilla-Silva 2004; Bonilla-Silva and Glover 2004) suggest that the United States should move from such a "U.S. style" of race relations to one recognizing a greater spectrum across racial and color lines, as is the case in Latin America. In much of Latin America, such as Brazil (Telles 2004), numerous terms are used to identify people racially and, unlike in the United States, people with dark skin of high socioeconomic standing may be viewed as white, as "money whitens." In the Dominican Republic, many deny their African roots in favor of their indigenous *Taino* and Spaniard roots (Jiménez 1996, 2006, 2008; Robles 2007; Stinchcomb 2004). In this setting, Dominicans with dark skin, who would be seen as black in the United States, are viewed as white. Indeed, the term black is reserved almost exclusively for Haitians. Dominicans and other Latin Americans with dark skin get a rude awakening when they come to the United States and are viewed as black.

We draw upon the works of scholars across many disciplines to ground our theoretical arguments regarding the role that race continues to occupy in the United States and in other parts of the Americas. As this essay illustrates, tracing the role of race in society requires scholars, including sociologists, to broaden their analytic gaze to incorporate both a wider geography and other academic disciplines. Furthermore, to begin to understand the enduring role of race in society requires rethinking the traditional methods of the social sciences. Not only have our political and social structures been constructed and defined by the dominant group, so, too, have the methods of the social sciences. For example, early sociologists such as Sumner, Ward, Park, and Cooley, among others, who laid the foundations upon which we still build, were heavily influenced by the racial stereotypes of their day (see Zuberi, 2001). For the contemporary study of race and racial stratification, this means unraveling these racist ties and looking anew at the (racial) assumptions that undergird much social science research (Zuberi and Bonilla-Silva, 2008).

In sum, we argue that not only does a color line continue to demarcate U.S. society, but also that it is so entrenched in our bureaucratic structure that further progress will take uncovering and acknowledging these linkages, publicly airing them, and ultimately unraveling them in order to construct a fairer, more just America. Moreover, many persons from other parts of the Americas immigrate to the United States, where they often encounter clashes between how they were viewed racially in their home country and how they are viewed in this country. Our broad approach, drawing from a variety of disciplines and understanding of racial matters across the Americas, provides insights for the teaching of race in the Americas.

A Brief Discussion on Race in America

Racial classifications and the designation of the U.S. population into racial categories stem from the beginning of the new union (Hochschild and Powell 2008). Indeed, the first U.S. census conducted in 1790 designated between "Free Whites," "All Other Free Persons," and "Slaves." In 1820 a new category for "Free Colored Persons" was introduced. The next decade saw a further parsing of the white population into "Naturalized" and "Not Naturalized" categories.

The first reference to mixed races appeared in the 1850 census when the population was categorized as "White," "Black," and "Mulatto."

"Chinese" and "Indian" were new racial categories added to the 1870 census forms. The 1890 census tried to more finely quantify the mixed-race population through the distinction of "quadroons" (one-quarter black) and "octoroons" (one-eighth black). Further, "Japanese" appeared for the first time as a racial category in the 1890 census, while terms for "Koreans," "Hindus," and "Filipinos" were added in 1910. The 1930 census witnessed "Mexican" as a new racial category.

Censuses throughout the twentieth century continued to adjust racial categories. A directive from the Office of Management and Budget (OMB) required that for the 1980 census race be designated into one of four categories—American Indian or Alaskan Native, Asian or Pacific Islander, Black, or White (see Snipp, 2003). A separate question determined Hispanic or non-Hispanic ethnicity. Furthermore, since the 1980 census, people have self-identified their race and Hispanic origin. Moreover, beginning in the 2000 census, people could designate more than one race, as established by the 1997 revisions of OMB Directive No. 15 (see Snipp, 2003).

In retrospect, U.S. society has had difficulty in classifying mixed-race individuals. As James (2008) points out, the social understanding is that mixed-race individuals are not white. As the inclusion of categories to deal with mixed-race persons on the census forms indicates, discomfort with a "white" label is most acute for mixed-race individuals who are part black. For these persons, the one-drop rule has traditionally placed them in the black category. Indeed, when a mixed-race person identifies as white, they are often understood to be "passing" (James, 2008).

The white racial category has been similarly protected. As James (2008) has argued, the ongoing issue of how to classify the mixed-race population reveals the long-standing concern with protecting the "white" category. For example, past census instructions required classifying multiracial children according to the race of the nonwhite parent. Further, the Census Bureau has long struggled with how to label the Latino population. Conventional wisdom recognized long ago that while the bureau classified Latinos as "white," Latinos are not "white." In fact, the 1930 census, for the only time, treated "Mexican" as a racial category. The 1970 census introduced the umbrella term "Hispanic" to include persons whose ancestors originated from Mexico, Puerto Rico, Cuba, the Dominican Republic, South or Central America, or other Spanish origin. Yet, the term "Non-Hispanic white" is part of the American racial taxonomy and "white" remains protected.

No doubt an adequate racial definition for classifying people living in the United States remains elusive. However, so central is the role of race in our society that we continue to try. This discussion illustrates the socially constructed nature of race and the variable meanings of race, which depend upon the historical context in which they appeared.

Immigration, Citizenship, and Race

Traditionally, citizenship has been conferred through birth or naturalization. The decisions regarding who and under what conditions a person can enter the United States and then, once here, the conditions necessary for citizenship are powerful ones. Indeed, that this activity has been regulated almost from the nation's founding reveals that the United States, far from being the beacon to the world's poor and oppressed, has been far less welcoming and democratic than the rhetoric indicates. While the Declaration of Independence declared that "all men are created equal," our history abounds with examples of contradictions and tensions regarding citizenship. Sociologically defined, citizenship is "a set of practices...which define a person as a 'competent' member of society...." (Turner, 1993:2). In short, citizenship is about membership with associated rights and responsibilities.

Determining "competent" from incompetent members of society, the rights and duties associated with this membership, and the conditions for the realization of this membership are processes that are steeped in racial thinking. Concerns about the type of people coming into the country are long-standing (Jasso, 1988). The first naturalization law passed by Congress dates to 1790 and specified that any free, white alien who had lived in the country for at least two years was eligible for citizenship, provided they could prove their moral worth, were not likely to become a public charge, and were willing to take an oath of allegiance. Thus, being white was a prerequisite for becoming American. In 1870, the naturalization law was amended to confer citizenship onto former slaves, thus establishing a black-white color line for determining eligibility for U.S. citizenship.

Immigration laws have likewise been fashioned using blatantly racist means. Race was the defining feature in decisions regarding who to include and who to exclude from the United States. As Ngai (1999, 2004) has shown, U.S. immigration laws drew color lines around the world, separating countries deemed "white" and "nonwhite," with only people from the former being eligible for immigration and naturalization. For example, the Chinese Exclusion Act of 1882 banned all immigration from China, with Japan added to the exclusion list through the Gentlemen's Agreement in 1908, and other Asian nations added in 1917 to the "barred Asiatic zone...." (Ngai, 2004:18). The 1924 Immigration Act, one of the most restrictive immigration laws in U.S.

history, created a hierarchy of desirability for the world's inhabitants with rising levels of restriction on the least desirable. The act specified a quota of 2 percent of the total of a given nation's residents as reported in the 1890 U.S. census (Daniels, 2004). Curiously, all nonwhite U.S. residents were excluded from definitions that determined the quotas for immigration (Daniels, 2004). Daniels concludes: "If anyone requires evidence that Congress regarded the United States as a 'white man's country,' this clause...provides it" (p. 55).

During this era, aside from former slaves who became U.S. citizens, persons deemed nonwhite were denied U.S. citizenship (Carbado, 2005). It was the job of the federal courts to determine whiteness and thus eligibility for U.S. citizenship. According to Ngai (2004), between 1887 and 1923 the federal courts heard twenty-five cases in which the racial status of those seeking citizenship was questioned. Further, the definitions of whiteness were neither static nor consistent, but vacillated between the science of eugenics and "common sense" (Lopez, 1996; Ngai, 2004). In two landmark decisions in 1922 and 1923, the U.S. Supreme Court ruled that Japanese and Asian Indians in the United States were not white and thus were ineligible for citizenship.

The easing of the blatantly racist provisions of immigration and naturalization laws began in the aftermath of World War II in the environment of wartime labor shortages and postwar economic prosperity. The McCarran-Walter Act of 1952 removed the most egregious racist and discriminatory elements of U.S. immigration and naturalization laws, including those barring Asians. The liberalization of immigration and naturalization laws begun in the 1950s was completed in the 1960s with the lifting of all quota restrictions.

The consequences of both the long-standing racist and discriminatory nature of the immigration and naturalization laws and their eventual liberalization would affect immigration to the United States in ways unanticipated at their crafting. As we detail next, today's immigrants and racial and ethnic minorities are a reflection of (and burdened by) our racist past.

Racial Legacy

For most of U.S. history, immigration, citizenship, and race were explicitly linked. Nonwhites were intentionally kept out of the United States, and citizenship was reserved for whites. While the definition of who was "white" has been contested, it nonetheless continued to be the defining feature in U.S. naturalization laws. The 150-plus years of overtly racist laws have molded the racial and ethnic lines that are still clearly demarcated in U.S. society today.

For example, the long legal link of whiteness and citizenship clearly established whiteness as the normative identity for U.S. citizenship (Carbado, 2005). Carbado (2005:637) notes that "Naturalization is not simply a formal process that produces American citizenship but also a social process that produces American racial identities." What does this mean for today's minorities? One consequence of this legal linking of whiteness and citizenship today is the social delinking of citizenship and American identity, particularly for those who are not white. As Carbado (2005) points out, most of the Japanese interred during World War II were not identified as "Americans" despite being American citizens. Indeed, the modern characterization of Asian Americans as "perpetual foreigners" points to the enduring legacy of these normative associations of identity, race, and citizenship.

The National Origins Act of 1924 restricted, but did not bar, immigration from Southern and Eastern Europe. Thus, white Southern and Eastern Europeans, although regarded as inferior to those from other parts of Europe, were still viewed as assimilable into American society. Hence, the act still conferred and affixed the "white" racial category to these groups thereby paving their way to U.S. citizenship. In contrast, nonwhites living in the United States were ineligible for U.S. citizenship. Excluded from the vision of "American" in the 1920s, these same groups remain somewhat peripheral to the vision of "American" today.

But what does it *mean* to be "American"? To be white, Anglo-Saxon, and Protestant (WASP); speak English; and reside in the United States rather than Canada or Mexico is the short answer. One of the definitions of "Anglo-Saxon" "a person who speaks English" (s.v., dictionary.com) Having discussed the white connection to the definition of American, we now turn to a discussion of the prominence of the English language to U.S. culture.

"Do You Speak American?"[1]

In an essay on language, Heritage (2006) describes language as the basis of social identity and culture. While a multitude of different languages is spoken in the United States, the vast majority of U.S. residents (80 percent) speaks only English at home. To "speak American," then, is to speak English.

However, it is not enough to simply speak English, but to speak *Standard* English. Recall the controversy that arose when the Oakland School Board in 1996 declared Ebonics (Black Urban Vernacular English) the official language of African American students who attended its schools. Lost in the uproar of this decision were the reasons given for this bold move—to address the language barrier that confronts many African American students and foster lines of communication between teacher and student. Nevertheless, in the maelstrom that followed, the Oakland School Board quickly retreated from this position.

Orlando Taylor, a linguist, claims that "Language is a reflection of a people" (Hamilton, 2005:35). According to Taylor, if a group of people are considered inferior, it follows that so, too, is their language. One thing that the ensuing uproar over the Ebonics debate made clear is that the language used by many African Americans is considered inferior.

Stereotypes regarding a person's intellect are often associated with the enunciation and pronunciation of English, as we saw in Barack Obama's presidential campaign. Recall Senator Joe Biden's description of Obama as "the first mainstream African-American who is articulate and bright and clean and a nice-looking guy. I mean, that's a storybook, man" (as quoted in Coates, 2007). While it is no secret that Obama is biracial and multicultural, this nuance was nowhere to be found in Biden's comment. Although "compliments" such as this one play out before the public, it does not take much to imagine what happens behind the scenes and away from microphones and cameras (see Picca and Feagin, 2007).

Clearly there is a stratification system that ranks regional dialects along a continuum anchored by Standard English on one side and black English on the other. Yet, speaking English with a foreign accent warrants a separate discussion. English as a second language, particularly when the first language is Spanish, is problematic to many in the United States. So strong is the mainstream attachment to English (and distaste for Spanish—the nation's second most common language) that attempts are made regularly to have English declared the country's official language. And because these efforts have failed at the national level, states have taken matters into their own hands. West Virginia, one of the states with the fewest Latinos, is seeking to adopt English as its official language. In fact, thirty states have made English their official language (*ProEnglish Action*, 2010).

Reinforcing the negative perception of both the Spanish language and the Spanish-speaking population are the legal cases in which judges have viewed a child's lack of English comprehension (and use of Spanish) as abuse on the part of the parents. For example, the *New York Times* reported that a Texas judge accused a mother of abusing her child for speaking to her only in Spanish and then threatened to remove the child from her custody unless the child began speaking English (Verhovek, 1995).

In short, language and more specifically standard English is much more than a vehicle for communication. The language one speaks is a symbol of identity, citizenship, patriotism, and apparently, even parenting skills. Thus, to have one's language challenged (or the ability to speak the dominant language) is to also question one's being on a variety of dimensions and to imply that one is of low social status.

"What Are You?"

Questions like "What are you?" are not uncommon for people whose appearance does not conform to the "American" image. Yet, one thing is clear just from the asking—the asker has precluded the option of American from the list of possible answers. Canadian citizen Marsha Giselle Henry's (2003) parents emigrated from Pakistan to Canada during the mid-1960s. Like many multiracials, Henry (2003) reveals she is constantly asked about her "pedigree" and feels uncertain about how she represents herself on a daily basis.

Sitting outside the mainstream seems to give license to continuous questioning of one's identity. How to answer questions such as "what are you?" and "where are you really from?" is difficult and anxiety provoking. Henry (2003:234) laments, "I remember feeling angry and frustrated when people challenged my representations, always probing and asking additional questions and then resigning themselves to some first impression." As Wu (2002) explains, people whose own American identity is assured do not seem to understand how offensive these questions are.

> Like many other people of color...who share memories of such encounters, I know what the question "Where are you *really* from?" means, even if the person asking it is oblivious and regardless of whether the person is aggressive about it. Once again, I have been mistaken for a foreigner or told I cannot be a real American. (80)

For multiracial individuals, the issue of identity is further confounded by inadequate and simplistic racial categories. To be multiracial in the United States is to have no home or to be a stranger in one's own home. In Barack Obama's bid for the Democratic Party's nomination for U.S. president, concerns were expressed within the black community as to whether Obama was indeed "black enough." That, in fact, was the title of a *Time* magazine article: "Is Obama Black Enough?" In the article, Coates (2007) notes that while Obama's biracial identity made him acceptable to many whites, it "also opened a gap for others to question his authenticity as a black man." Despite African Americans such as Stanley Crouch questioning his blackness, Obama succinctly illustrates how he is viewed in the larger society: "If I'm outside your building trying to catch a cab," he told Charlie Rose, "they're not saying, 'Oh, there's a mixed race guy.'" (Coates 2007).

Similarly, professional golfer Tiger Woods was roundly criticized by many blacks for his self-identification as "cablinasian," a term he coined to encompass his mixed-racial ancestry (Kamiya, 1997). Such is the plight of multiracial persons who are neither totally accepted nor rejected and are defined by others as they see fit, especially if one refuses to answer the series of questions: "What are you?" "Where are you *really* from?" "Where are your parents from?" meant to locate one into a racial category so as to ease the anxiety of the asker. Per Wu (2002): "they have placed me in their geography of race and somehow know all they need to know. They must feel they have gleaned an insight into me by knowing where I am '*really*' from. They can fit me into their world order" (Wu, 2002:81). As Pearl Fuyo Gaskins (1999) notes, the question "what are you?" indicates that far from not seeing race, instead, it is the first thing we notice about a person. And when a racial identity is not readily discernible, we are thrown into a kind of crisis that is "caused by the contradictions between how people have been trained to understand race and the fact that the multiracial person doesn't fit that scheme" (pp. 20–21). In many ways, then, people whose American membership is questioned are placed in a subordinate position vis-à-vis the mainstream.

Per Blumer (1965), dominant groups work to maintain, reinforce, and protect their privileged group position by way of the color line. Power imbalances between dominant and subordinate groups contribute to the maintenance of the color line. So, too, do assimilation ideologies. As Douglas (2008) articulates, one problem with assimilationist assumptions is the superficiality that an American can be gleaned using a narrow spectrum of physical appearance. Alba and Nee (2003) define assimilation as the "attenuation of distinctions based on ethnic origins" (p. 38). But what does it mean to "attenuate" one's ethnic/racial distinctions? Per Alba and Nee (2003), "a key to assimilation...is boundary spanning and altering" (p. 59). Boundary crossing can also occur across generations through intermarriage, which ultimately results in "...individuals...whose social appearance is indistinguishable from majority group members of the same social class and region" (Alba and Nee, 2003:61). As shown earlier, assimilation as Alba and Nee define it is an impossibility for a large swath of the U.S. population, who sit outside the "American" mainstream. Further, this narrow image does not allow for or acknowledge the increasing multiracial population.

In sum, as Douglas (2008) argues, assimilationist theories and assumptions fail to grant that the immigrant experience (and the assimilation process) is one that is largely defined by the dominant group. Persons outside the mainstream are situated and defined relative to the dominant group whose interests are in protecting both their status and higher position within the stratification system. It is to illustrate this point that we turn to next.

The Dynamics of Privilege, Affirmative Action, and Retrenchment

Groups in a position of power use all means at their disposal to maintain their status and to keep competing groups at a disadvantage. This perspective, extending back to Karl Marx is used to recognize how whites have kept their superior economic, social, and political position vis-à-vis groups of color. As noted earlier, this was done originally through the definition of citizenship drawn on the basis of whiteness, which provided access to societal resources. The racial hierarchy has been sustained by the benefits that whites gain from the existing stratification system and the historic subordination of nonwhites (Bonilla-Silva 2009; Feagin 2006; Jensen 2005). For all of U.S. history, whites have not only dominated in terms of power and resources, but also demographically.

In this setting, whites have been the beneficiaries of white privilege. More specifically, when societal resources have been marshaled to deal with an economic crisis or to create social and economic opportunities, they have been unequally placed in white hands. Thus, whites were the disproportionate beneficiaries of massive social programs such as New Deal programs and the G.I. Bill (Katznelson, 2005). Social benefits stemming from these programs, which opened doors to higher education and homeownership, were allotted at the local and state levels, with the white power structure routing benefits to whites. In essence, in the words of Katznelson (2005), whites were the beneficiaries of affirmative action long before formal affirmative action programs to address racial inequities were created. Whites benefited from and gained access to societal resources and social, economic, and political opportunities simply by being white. Even when the War on Poverty was initiated in 1964, the majority of the poor was white, making up 71 percent of the nation's poor in 1960.

The period between 1954 and 1964 saw the development of policies designed to provide minorities access to the opportunity structure (Saenz et al., 2007). The Supreme Court's 1954 ruling in *Brown v. Board of Education* along with the passage of the Civil Rights Act of 1964—which called for, respectively, the integration of public schools and opened up opportunities for minorities in such areas as voting, housing, and education—both speak to important policy developments during this era. Yet, in the context of groups in power using all means at their disposal to hold their power and privilege, individual and structural efforts were made to dampen the impact civil rights rulings and legislation. Thus, many educational institutions in the South fought viciously to bar blacks from gaining entrance. Moreover, many white families subverted the desegregation of education by voting with their wallets and feet, and placing their children in private schools or moving to places with few persons of color. In addition, the principles of civil rights and affirmative action laws were undermined through ideology and the political system. For example, many whites cried "foul" when such legislation was put in force, claiming that they were victims of "reverse racism" or "reverse discrimination." Beginning with the *Bakke* decision of 1978 and a series of court cases related to desegregation of public schools in the K–12 system, a tremendous reversal of policies designed to allow minorities access to the societal opportunity structure was well underway by 1980 (Orfield and Eaton, 1997).

The post-civil rights era has coincided with a major demographic shift that has changed the face of the United States, with the share of whites declining significantly. As people of color—especially immigrants from many of the nations of the Americas—are increasingly part of the ranks of youth and the poor, we have seen an unraveling of the safety net, starting in the 1980s. The large investments that we saw in the post-World War II era, which benefitted whites disproportionately when they were a large share of the youth and poor population, have not been marshaled over the last few decades to allow today's youth and the poor to gain access to the opportunity structure. Latinos, the nation's fastest-growing group, continue to have extremely high dropout rates, with one in two dropping out of high school in many areas (Velez and Saenz, 2001). We suspect that dropout rates of such magnitude would be intolerable and commissions and policies would be put in place to deal with this issue if the group in question was white rather than Latino.

The Changing Demography of the U.S. Population

While a black-white paradigm has been used historically to understand race relations in this country, the reality is that the white population has declined in a relative sense and the black population has remained fairly stable. Indeed, major changes in the U.S. population have been afoot for decades resulting in dramatic changes in the racial and ethnic tapestry of the population.

Overall, the U.S. total population increased by 8 percent between 2000 and 2008. The non-Hispanic white (single-race) population grew much more slowly (2.3 percent), while the black population growth rate (8.1 percent) mirrored the national growth level. However, the growth in the Latino (32.8 percent) and Asian (30.8 percent) populations was especially impressive, each expanding about fourteen times faster than the white population and about four times faster than the black population.

It is obvious that Latinos represent the engine of the U.S. population change. Latinos accounted for one of every two persons added to the national population between 2000 and 2008. Furthermore, Latinos and Asians—the two groups growing the most rapidly—comprised two of every three persons added to the U.S. population during this period. The varying growth across racial and ethnic groups has led to changes in the relative size across these lines. For instance, the percentage of the U.S. population that is Latino increased from 12.5 percent in 2000 to 15.3 percent in 2008, with the percentage of Asians also is rising. The relative size of African Americans remained stable at 12.1 percent in the two time periods. The share of people in the country who are white dropped from 69.1 percent in 2000 to 65.4 percent in 2008 (it was 80 percent in 1980).

We are increasingly moving to a minority-majority country where nonwhites will be the numerical majority. This is already happening among the youth in a variety of states across the country, a pattern that portends future changes in the racial and ethnic composition of the country as these young people age. Of all babies born in the United States in 2006, 46 percent were nonwhite (Martin et al., 2009). As these youth age, they will contribute to the rise of minority-majority populations across these areas.

Population projections conducted by the U.S. Census Bureau (2009) document the deepening racial and ethnic transformation that the United States will experience in the coming decades. The U.S. population is projected to increase from 308 million in 2010, to 358 million in 2030, and to 399 million in 2050. While each nonwhite racial and ethnic group is projected to grow between 2010 and 2050, this is not the case with whites. The white population is expected to dip between 2010 and 2050, with the population decline expected to occur between 2030 and 2035.

Although all nonwhite populations are estimated to increase between 2010 and 2050, the growth is concentrated among Latinos and to a certain extent Asians. The Latino population is projected to more than double from 48.5 million in 2010 to 110.7 million in 2050. Moreover, the Asian population is expected to expand by 77 percent, from 13.3 million in 2010 to 23.6 million in 2050. African Americans and American Indians and Alaska Natives are projected to grow at a relatively slower pace.

Accordingly, the U.S. population growth will increasingly depend on Latinos. Of the 90.6 million persons that are projected to be added to the nation's population between 2010 and 2050, two of every three are expected to be Latino, reflecting an increase from 16 percent in 2010 to 28 percent in 2050. In contrast, the share of the white population in the U.S. population is expected to decline progressively, from 65.2 percent in 2010 to 49.9 percent in 2050.

Thus, the disproportionate growth of the Latino and Asian populations alongside the decline in the white population and the stability of the black population is placing challenges on the black-white framework. Other demographic changes are in progress that also call for the expansion of the black-white framework.

Major transformations in intermarriage have occurred over the last few decades, with concomitant changes in the way that the offspring of such unions view themselves racially and ethnically. In 1980 about one of every thirty-one marriages was exogamous (spouses of different racial/ethnic groups); by 2008 approximately one of every twelve marriages was exogamous. There are several variations in the prevalence of exogamy across racial and ethnic groups. For example, whites and blacks cross racial or ethnic boundaries the least, a reflection of the rigid color line that continues to define race relations in this country. Yet, the greatest increases in exogamy between 1980 and 2008 have occurred among whites and blacks. In relative terms, whites and blacks were about three times more likely to cross racial lines in marriage in 2008 than they were to do so in 1980.

The increase in intermarriage over time has led to the rise of offspring who identify with more than one race. The 2000 census for the first time allowed people to classify themselves as multiracial. It tallied about 6.8 million multiracial persons, about 2.4 percent of the nation's population. The number of multiracial persons rose, albeit slowly, to 7 million in 2008. The demography of the multiracial group, however, suggests that the multiracial population will increase noticeably in the future. In particular, multiracial individuals are a much younger population than single-race individuals—the median age of the multiracial population was 18 in 2008 compared to 37 for the single-race population. Given the youthfulness of multiracial individuals, it is no surprise that population projections suggest that the multiracial population will increase significantly over the coming decades. The projections suggest that the multiracial population is expected to nearly triple from 5.5 million in 2010 to 15.4 million in 2050.

In sum, the twenty-first century will increasingly result in greater racial and ethnic diversity in the United States. The nation's population will be increasingly Latino and Asian and less white. Multiracial individuals will become a growing part of the national racial and ethnic tapestry. These changes then place major challenges to the existing black-white paradigm that has guided our understanding of race relations in this country. It is apparent that today many Americans do not know where Latinos, Asians, and multiracial persons fit into the racial and ethnic landscape.

Challenges to the Existing Black-White Paradigm

These demographic changes deviate from the traditional black-white paradigm that has been used throughout U.S. history to understand racial and ethnic matters. The rigid dichotomous color line in the United States has been quite influential in the way we think about race and ethnicity, how we define and measure race and ethnicity, the construction of policies related to race and ethnicity, and the development of theoretical perspectives for understanding race and ethnicity. Latinos, Asians, Native Americans are often squeezed into models and perspectives that continue to try to place them into the established "white" or "black" categories.

The practice of forcing racial and ethnic groups that are not black or white into one of the two dichotomies reflects the lack of appreciation or understanding of the specific histories and contexts in which these groups have existed in the United States. In line with the benefits of white privilege, the history of the United States that we learn throughout grade school and into high school is *white* history. The history of groups of color tends to be packaged into materials related to the civil rights movements and civil rights policies, with the greatest emphasis on the black population. There is far less concern and understanding regarding the unique histories of the multitude of groups that comprise the Native American, Asian, and Latino populations.

Further, immigration matters complicate the way many Americans view the country's two fastest-growing racial and ethnic groups—Latinos and Asians. Despite their long presence in the United States, Latinos and Asians continue to be viewed as "perpetual foreigners"—people who do not fit into the American image. Among Latinos, Mexican Americans were initially incorporated into the United States in 1848 with the signing of the Treaty of Guadalupe Hidalgo following the end of the Mexican-American War. Thus, many persons of Mexican origin can trace their roots to the United States for four, five, six, or more generations. Among Asians, Chinese, Japanese, and Filipinos began immigrating to this country in the nineteenth century. Thus, many Asians from these groups can also trace their roots to the United States for many generations. Still, Latinos and Asians continue to be viewed as foreigners.

Consistent with the "foreigner" conception of Latinos and Asians in the United States, these groups are absent in many areas of life in this country. In the mass media, we see few Latinos and Asians on the Hollywood screen, in television sitcom shows, as newscasters, and even in talk shows that try to bring in the "minority" perspective. Portales (2000) aptly shows how Latinos have been ignored in many dimensions of social life:

> Hispanics...have traditionally been ignored and continue to be disregarded, in spite of considerable talk about the professed value of "cultural diversity" and "multiculturalism." Indeed, if we pay close attention, we will observe that Hispanics are oddly missing from or only tangentially included in most dialogues about diversity, even when the needs of "minorities" are being discussed. When referencing Hispanics what we have, and one cannot soften the nature of the "oversight" much, is a vast and pervasive national unwillingness to acknowledge almost everything that is Hispanic.... (pg. 54).

In sum, Latinos and Asians continue to be viewed as perpetual foreigners—people who do not fit the American image. As these two major groups play an increasingly important role in the shifting demography of the U.S. population, will this image change? Will they be seen as truly American? Will the definition of what constitutes American change?

Concluding Remarks

So what to make of all of this? One take is provided by Cara Lockwood (2005), author of *Dixieland Sushi*, when asked the meaning of "American":

> I think there is not one definition of an American. Right now there's a lot of discussion about how divided we are as a nation politically, but I

hope that we don't forget that one of the greatest things about our country is that there is room for so many combinations of cultures and attitudes.... I hope that as a nation we continue to value our diversity because I think it's our strongest attribute as a nation. The definition of being American is that there is no one definition.... (287)

This is a hopeful vision. As this essay has detailed, we as a society have not exactly embraced our diversity—indeed, we have done much to deny its existence. Racial and ethnic inequalities continue. Unfortunately color-blind rhetoric minimizes this reality. Pointing this out subjects one to charges of whining and/or playing the "race card." This is not constructive nor does it acknowledge or begin to grapple with the racial legacy that still lives in this country.

As the history of the United States shows, citizenship has been a contested asset that has not been conferred easily. For long, only white immigrants were eligible for U.S. citizenship. We continue to have these debates now as the nation grapples with the issue of immigration reform. Much debate has focused on what to do with the 13 million or so undocumented immigrants in the country, particularly when it comes to the establishment of a path toward citizenship. In the post-9/11 environment associated with high-tech security, massive worksite raids, the criminalization of immigrants, and the deportation of immigrants including naturalized citizens, it is clear that citizenship status is a valuable commodity. Its absence places people in vulnerable situations politically, economically, socially, and legally. In particular, the absence of U.S. citizenship status has increased the vulnerability of Latino noncitizens to exploitation, fear, and the violation of basic human rights. Massey (2008) has suggested that Latinos, especially Mexicans, represent the new blacks due to the lack of citizenship among many members of this group. Thus, simple immigration laws prevent many from becoming Americans—at least on paper. The case of undocumented immigrants who came to the United States with their parents at a very early age—the 1.5 generation—and who have grown up and been educated and socialized only in this country illustrates the absurdities associated with the disconnection between the country that people know as "home" and U.S. citizenship status. Many such individuals continue to be deported to countries where they were born but for which they have little familiarity and connection. Many of these persons fail to fit into their countries of birth, and some experience hostility for not speaking the native language or adhering to the local cultural norms. In the case of 1.5-generation Mexican-origin youth who have been deported to Mexico, often the locals view these individuals as "Americans" and not as genuine

Mexicans. As in the case of race and ethnicity, it is clear that an American identity is itself a social construction.

The United States is part of the Americas consisting of North America, Central America, South America, and the Caribbean. Technically, persons from all of these regions are American. Yet, it has been the United States that has appropriated the American identity. To be American is synonymous with being from the United States. Imagine if a European country—say Germany—co-opted the European identity and that European became interchangeable with being from Germany. The second author of this article grew up in the Rio Grande Valley in South Texas, a predominantly Mexican American area. In the dynamics of race relations in the region, whites viewed Mexican Americans as "Mexicans" not as "Americans." Mexican Americans, too, referred to themselves as "Mexican" and to whites as "Americans." Thus, for many Mexican Americans from this area, as well as from many other parts of the country, to be called an "American" continues to sound foreign and unexpected and not inclusive of themselves.

While we appreciate the larger point that Massey makes referring to Latinos as "the new blacks," it reinforces our larger point that a broader racial vocabulary is needed to deal with the new racial reality. Indeed, we can learn from our American neighbors to the north and south. We agree with the contentions of Bonilla-Silva and his colleagues (Bonilla-Silva 2004; Bonilla-Silva and Glover 2004) and Wu (2002) that the United States should evolve from its traditional black-white model to one that recognizes nuances that go beyond the dichotomous black and white categories. The demographic shifts that we have described previously will likely push the country to change the way in which it has historically viewed racial and ethnic matters. Further, we can appreciate Canada's embracing of multiculturalism and its adoption of multiculturalism as a matter of national policy.

Nonetheless, it is important to recognize that even in Latin American countries where race is fluid and dynamic and where money can whiten people racially, individuals with darker skin and those with indigenous features continue to occupy the bottom rungs of the stratification system while persons with lighter skin and European features hold the higher levels of the stratification system (Campos, 2009; Murga, 2008). Likewise, in Canada, despite its celebration of multiculturalism, "visible minorities" (Asians, blacks, and Latinos) live in segregated neighborhoods with lower incomes compared to people of European ancestry (Fong and Wilkes, 2003). Furthermore, Canada's aboriginal and black populations account for a disproportionate share of Canadian prisoners (Roberts and Doob, 1997).

As we have demonstrated throughout this essay, the Americas in the most expansive sense of the word are far from color-blind. Indeed, race and racial hierarchies infiltrate every aspect of life from where we live, work, and play. All the same, we stand to learn from our brethren from throughout the hemishpehre and thus expand our own racial categories beyond black and white. Integrating Latinos, Asians, and other racial and ethnic groups into the white-black paradigm need not be a zero-sum game. As Wu (2002) asserts, Latino and Asian perspectives supplement rather than replace other perspectives. No doubt the demography of the United States has changed. A different lens is needed.

Our essay contributes to the teaching of the Americas in several ways. For example, our essay points to the importance of considering the historical past in understanding contemporary issues related to the social and economic standing of racial and ethnic groups. In addition, our essay points to the importance in considering insights from multiple disciplines in understanding the historical past and contemporary conditions of racial and ethnic groups. Moreover, our essay calls for the expansion of the lens from which we have traditionally viewed racial and ethnic relations in the United States and beyond. In particular, the growth of the Latino, Asian, intermarried, and multiracial populations will require the expansion of the traditional black-white paradigm to one featuring a greater degree of racial and ethnic spectrum as found in the rest of the Americas. Similarly, we call for the need to understand the American identity beyond the mainstream of the United States to encompass groups that have been seen as perpetual foreigners in the United States as well as people in the rest of the Americas.

Note

1. In January 2005, PBS aired a documentary entitled *Do You Speak American?* in which former NewsHour anchor Robert MacNeil journeyed across the United States seeking answers to this question. Despite the diversity of possible answers that the provocative question implies, MacNeil's stated purpose was to explore in detail American English. Thus, to speak "American" is to speak English.

CHAPTER 8

Interdisciplinary Approaches to Teaching the History of the Western Hemisphere

Moramay López-Alonso

In 1976, social historian Alfred W. Crosby Jr. published a book that offered a new and fresh perspective on the study of the history of the Western Hemisphere: *The Columbian Exchange*. This work sheds light on a different facet of the discovery of the American continent by Europeans. In proposing that Columbus's expeditions had a biological component, Crosby presents the environmental consequences of 1492. In it we learn that some of the features that transformed the American continent into what we know it as today were much influenced by factors of geography, demography, and biodiversity as much as by those of politics. Up until 1492, the peoples of the Western Hemisphere had basically existed in isolation from the rest of mankind. The arrival of European colonizers and their African companions had profound consequences that went beyond the elements of colonization as traditionally studied by historians. The consequences were the by-product of the exchange of pathogens, plants, and animals. Diseases brought from Europe were a big killer of Indians. Measles and smallpox were more lethal than any gang of conquistadors in the battlefield.[1] In return, the Americas gave colonizers venereal syphilis; the disease was endemic in the New World, but in sixteenth-century Europe, it spread as a lethal epidemic.[2] Syphilis proved to be a rapid traveler, and by the end of the 1490s it had traveled from London to Moscow.[3]

This explanation of the exchange of diseases belies the traditional Black Legend of the Spanish conquest minted by sixteenth-century chronicles alleging that the violence of the conquistadors was the main cause of the decimation of the native population of the Americas. But Crosby's argument did not intend to contradict the notion that conquistadors were violent towards the aborigines; rather, it clarified that the biological encounter of both worlds was more powerful a weapon than any army. In the book's preface, Crosby discusses the importance of using multidisciplinary tools when teaching history as well as the relevance of including findings drawn from other disciplines:

> Before the historian can judge wisely the political skills of human groups or the strength of their economies or the meaning of their literatures, he must first know how successful their member human beings were at staying alive and reproducing themselves... [I] am the first to appreciate that historians, geologists, anthropologists, zoologists, botanists, and demographers will see me as an amateur in their particular fields. I anticipate their criticism by agreeing with them in part and replying that, although the Renaissance is long past, there is great need for Renaissance-style attempts at pulling together the discoveries of the specialists to learn what we know, in general[,] about life on this planet.[4]

Crosby's book was written more than thirty-five years ago, but its approach remains an important one for at least two reasons: First, over the last thirty-five years there have been advances in the natural and social sciences and new research technologies that facilitate the incorporation of their techniques and findings into the discipline of history. Second, the growing awareness of the need to preserve the environment has made environmental history a relevant subject in the curricula of the humanities and social studies.

In adopting Crosby's perspective to teaching history, the American continent can be conceived as a geographic unity. In this unity there is much to be learned from studying climatology, demography, language, and culture. This exercise can be accomplished through the use of multidisciplinary tools in a way that each discipline can contribute its reading of a given subject, while the combination of all readings can provide a more complete answer to historical questions. Such an approach complements and enriches the traditional curriculum of modern history—work that often highlights the rise of nationalism and the formation of political entities that comprise today's countries in the Western Hemisphere.

In this essay I present a concrete example of the multidisciplinary approach to teaching history using the Western Hemisphere as a primary subject of study. I use a historical event that purports to be a breakthrough moment in the history of the Americas: the demographic transformation of the first century that followed the arrival of Europeans in the Western Hemisphere. I show how by including findings from disciplines outside of traditional history, we can gain a more integral perspective in our understanding of the major events and their consequences. At the end of this exercise I will assess the challenges and benefits of adopting such a perspective.

The Demographic Consequences of the Arrival of Europeans into the New World

Few episodes in the history of humankind have been as dramatic in their repercussions as the arrival of Europeans to the American continent beginning in 1492. Isolated from other continents, the Western Hemisphere had distinctive inhabitants, flora, and fauna.[5] The penetration of Europeans into the New World altered almost every aspect of life in this continent.[6]

The chronicles that narrate the encounters of the Old and New Worlds date to the first years after Christopher Columbus's arrival in 1492. They were written with different purposes, such as to inform the Spanish royal authorities—who sponsored the first ventures—on the findings of the expeditions; to report on progress made in the conquest and evangelization projects; to justify the moral, religious, political, and economic validity of colonization; or to document a travel journey to share discoveries with people back in Europe. Whatever the motivation behind the creation of these chronicles, a modern reading of these sources allows us to uncover the different aspects of life in the Western Hemisphere that were transformed as a result of the first encounters.

Of the earliest chronicles, the sixteenth-century works of Fray Toribio de Benavente and Fray Bartolomé de las Casas are today perhaps the most widely read on the subject of the conquests. Bartolomé de Las Casas was born in Seville in 1484, traveled to Santo Domingo in 1502, and himself became an owner of Indian slaves not long thereafter. In 1510, he became a secular priest, and by 1522, he joined the Dominicans. It is then that he acquired a principled awareness of the Indians' suffering and plight.[7] Las Casas was the creator of the Black Legend of the conquest of the New World. He began writing prolifically with the purpose of denouncing the abuses that European colonizers were inflicting on the indigenous populations. He argued against the exploitation and enslavement of Indians. The last decades of his life were devoted to the quest to end these abuses, and earned him the title of "Defender and Apostle of the Indians" and turned him into a pioneer of human rights. Las Casas realized that as a priest and evangelizer, his most powerful tool was his writings. His works were not a mere intellectual exercise or an attempt to write the history of the conquest of the New World, and his works are not outstanding for the quality of their prose. Rather, they were intended to reach large audiences, and "composed to make the simple naked truth not merely evident but also compelling."[8] His objective was to attract the attention of the Spanish royal authorities, for he believed they could put an end to colonizers' abuses. To make his arguments, Las Casas highlights the excesses of the conquest and its consequences. But as a result his work is rife with exaggerations and imprecision; for instance, he calculates the number of deaths in Hispaniola over the first forty years of the conquest at 12 million, and he adds that his estimates might be underestimated.[9] Despite the value of being one of the first and main primary sources available for this chapter of history, the content of Las Casas's works must be taken with reservations by historians who value the accuracy of primary sources. Although not a thorough scholar, Las Casas did prove to be a successful communicator: His writings were widely circulated, and the Spanish monarchs Charles V and Philip II heard his claims—their policies toward the Amerindian population were influenced by Las Casas advocacy. At the same time, Las Casas's *Short Account of the Destruction of the Indies* was also used by rival monarchies in Europe in the propaganda wars against the Spanish and Habsburg empires, as well as Catholicism, as the Protestant reformist movement was gaining strength.[10]

Fray Toribio de Benavente, also known as Motolinía, like Las Casas, was a member of a religious order. One of the twelve Franciscans who arrived in the wake of Hernán Cortés's conquest of Mexico, he proved to be much more than a proselytizer of the Catholic faith among the indigenous population.[11] In contrast to Las Casas, Motolinía's scholarship—which precedes Las Casas's works—was only known to a few royal authorities and some of his close collaborators, and was not published until three centuries after his death.[12] Still, Motolinía's works are a major source for later studies written by other scholars.[13] His academic training in Salamanca, Spain, included philology, which he studied with Antonio de Nebrija, the creator of the first grammar for the Spanish language. Motolinía's training in philology facilitated his learning of Nahuatl, the most widely spoken indigenous language in Mesoamerica. He was thus able to better communicate with and understand the indigenous population. He was also an advocate of the rights of Indians, but the scope of his works was wider than that of Las Casas's. Motolinía was more thorough in his scholarship, and he attempted to write a history of the colonization of Indies as well as a history of the peoples of Mesoamerica based on his research. His work is a forerunner in the field of ethnography, although he wrote with the purpose of justifying Spain's colonization of the New World in relation to the evangelization of the Amerindian population. To build his arguments, he relied on biblical comparisons, an approach that has led modern scholars to challenge the veracity of his chronicles.[14]

Leaving aside the differences of Las Casas's and Motolinía's works in depth and scope, they both coincide on a fundamental point: During the first decades of the colonization period, the number of deaths among the Indian population was alarming. In their eyes, Indian mortality due to disease, labor exploitation, and mistreatment all originated in the greed of the Spanish conquistadors.[15]

The content of Motolinía's and Las Casas's works was not challenged during the three centuries of Spanish colonialism. If anything, during the early part of the nineteenth century, as former Spanish American colonies gained independence, their works met with renewed appreciation, as the Black Legend was used by American insurgents to justify the necessity of independence after three centuries of colonial domination.[16]

Neither Las Casas nor Motolinía lived long enough to witness in all its harshness the epidemics of the second half of the sixteenth century and the resulting demise of the population. By the seventeenth century the Indian population had reached such a low number that it became a source of concern for Spaniards.[17] Centuries had to pass and scientific advances had to be made before scholars could aspire to estimate more accurately the size of the Indian population on the eve of the arrival of Europeans and investigate in more depth the causes behind the high mortality of the first century of Spanish rule.

In the twentieth century, with the development of new research techniques in the natural and social sciences, it became possible to investigate more systematically the different elements that contributed to the transformation of the Americas during the sixteenth century. As a point of departure, there was a consensus that the sixteenth century spanned a demographic disaster, especially in Mesoamerica, one of the most densely populated regions in the American continent.[18]

Before calculating the real magnitude of the decline in population, one question that underlay the first twentieth-century round of revisions of the encounter of the New and Old Worlds concerned the actual size of the population of the Americas at the time of the arrival of Europeans. Demographers by the mid-twentieth century had developed the research techniques to estimate the demographic history of Mexico prior to the second half of the nineteenth century, when the recording of vital statistics began.[19] In particular, Sherburn Cook and Woodrow Borah from the University of California, Berkeley devoted decades of their academic careers to this project.[20] As a starting point, they used the different chronicles of the sixteenth century; however, the accounts in these sources were vague and differed too greatly among themselves to enable statistical calculations that could yield reliable results. Therefore, Borah and Cook turned to tax and tribute records, complementing these with military and religious records to make estimations that yielded statistically significant results.[21] Cook and Borah thus produced the first statistical estimate of the size of the population of central Mexico at the time of the arrival of Hernán Cortés to Tenochtitlán (Mexico City) and calculated the ensuing decline in population. According to their calculations there were "approximately 16,870,000 persons alive in 1532 and 1,370,000 in 1595."[22] Later, by applying the same methodology they made calculations for the demographic history and evolution of other parts of the Spanish American colonies.[23]

Cook and Borah's estimation of the population generated a debate about its accuracy. It was suggested that their calculations overestimated the number of inhabitants of the central valley of Mexico in the first half of the sixteenth century. Three schools of interpretation emerged from this debate, each defending a point of view with regard to the size of the population on the eve of the arrival of Europeans to the Western Hemisphere and the magnitude of the subsequent demographic disaster: catastrophists, moderates, and minimalists.[24] Catastrophists placed the scale of demographic decline at 90 percent and a large native population at contact of 10, 20, or even 30 million. Moderates argued that the decrease was only 50 to 85 percent with a population at contact of 5 to 10 million, but agreed with catastrophists on population totals at the nadir (1 to 1.5 million during the first half of the seventeenth century). Minimalists consider the scale of disaster to be approximately 25 percent and the population at contact to be 4.5 million.[25] No school of interpretation won this debate, but there was an agreement on the fact that it was difficult to make an accurate estimation because all calculations were made using tax and tribute records rather than population censuses (no censuses were conducted then). In addition, there was the issue that, over this period, settlement patterns changed, making it difficult to do a follow-up over time. Changes in settlement patterns were due to relocation of communities, or because some communities changed names while others just ceased to exist due to disease, warfare, or ecological distress, as we will discuss later in more detail.[26]

There were several valuable gains in knowledge due to this debate: The extant sources were thoroughly revised and the methods of analysis tested in depth; researchers from other disciplines also offered valuable approaches. For instance, archeologists specializing in the emerging field of biorarcheology assessed population size based on their evidence of sites where they conducted research. Their findings show that there were some parts of Mesoamerica, such as Teotihuacán and Tenochtitlán, or certain parts of the Yucatán peninsula where densities of population reached numbers that were elevated for the Paleolithic era.[27] Still, these were exceptional cases rather than the norm within the sedentary regions of Mesoamerica. And these populations reached a peak in growth and later a decline due to different factors, some exogenous, such as climatic disorder and seismic activity, and some endogenous, such as population pressure, economic decay, or political disintegration.[28]

The calculations for the population and depopulation of the Americas in the sixteenth century as a result of the conquest also led to new research on the history of disease. Moreover, with advances in the field of epidemiology it became possible to better understand how diseases spread and why their intensity could diverge so much from one place to the other. It is around the mid-1920s that the concept of virgin soil epidemics was minted.[29] This concept involves the idea that when:

> a disease is introduced among populations which never before had experienced the illness, or which had been free from it for so long that any acquired immunity had disappeared, the populations will be attacked in a massive onslaught, with no age group or sex spared infection. Mortality, in consequence, will be high, perhaps even total. Such diseases appear with extraordinary virulence and at times exhibit symptoms quite unlike the ones we normally associate with them. After a period of acclimatization, which may last three or four generations, say eighty to one hundred years, a virgin soil epidemic will usually settle down to a somewhat milder form with the symptoms that were are accustomed to.[30]

Crosby applies this line of research to his work on the biological encounter of the New World, arguing that "[T]he fatal diseases of the Old World killed more effectively in the New, and the comparatively benign diseases of the Old World turned killer in the New... [T]he most spectacular period of mortality among the American Indians occurred during the first hundred years of contact with Europeans and Africans."[31]

Crosby argues that germs were at least as lethal as mistreatment and overexploitation of Indians.

Since the publication of Cook and Borah's and Crosby's first works, new studies on the history of disease and epidemics in the New World have been produced by researchers from various disciplines that offer a more detailed account of the nature of the different epidemics and that advance new hypotheses. Thus, advances in the field of epidemiology have provided us with a better understanding of the processes by which diseases are spread, the different mutations that pathogens might undergo, and the regularity with which epidemics can revisit a population. This knowledge also makes it possible to clarify certain events in history. There are at least three examples of this in the history of the Americas over the sixteenth century that are presented here: smallpox in Mexico, smallpox in Peru, and syphilis as a disease of the New World.

The Role of Smallpox in the Demographic Collapse of Mesoamerica

Sixteenth-century chronicles describe smallpox as the plague that caused massive death in Mexico. Smallpox is the quintessential example of a virgin soil epidemic. With the emergence of the debate on the magnitude of the demographic consequences of the arrival of Europeans to the New World, researchers began to question if smallpox had been the big killer sixteenth-century chronicles alleged it was; doubts continued as more knowledge on epidemics became available, and coincided with the celebrations of the fifth centennial of Columbus's arrival in the American continent.[32] It is worth highlighting that this revisiting of the conquest of the Americas represented Christopher Columbus's voyages as European intrusion. Against the view that Europeans had come to the New World to bring civilization, the new reading of this episode attempted to assess the magnitude of damage that Western civilization had inflicted in this continent. It should then not come as a surprise that the chronicles of Motolinía and Las Casas were re-read with a critical eye. Part of the criticism was based on the fact that the original chronicles made by Motolinía and Las Casas exaggerated the size of the Indian population at the time of the conquest as well as the number of deaths that followed. Another criticism of their work noted the vagueness with which they described the symptoms of the plague, which made their accounts even more dubious.[33] This revision of the conquest of Mexico also criticized the numerical calculation of the evolution of the population of Mexico by Cook and Borah, arguing that it was too high and that it gave too much importance to the 1520s smallpox epidemic, given that there were other epidemic outbreaks throughout the century in the 1540s and in the 1570s that were just as deadly as the one observed in 1520.[34]

This new interpretation of the encounter between the Old and New Worlds encompassed a new appreciation for the viewpoints of the colonized. As mentioned earlier, Motolinía, along with many of the first evangelizers, was a trained philologist who was able to learn indigenous languages and created grammars for some of the most widely spoken indigenous languages of the Americas. Motolinía and these peers also taught indigenous scribes how to write their language in the Roman alphabet. This training made it possible for the scribes to write documents in their language in a readable format for those who knew this language. At the end of twentieth century, appreciation for the indigenous perspective has led to the study of these indigenous colonial documents; their contents contrasted with those of Spanish language sources.[35]

In response to those scholars who minimized the importance of smallpox as a big killer epidemic in Mexico, claiming that they were based on exaggerations in sixteenth-century Spanish chronicles, other scholars used indigenous language sources and compared them to Spanish-language ones. For the study of the sixteenth century, this proved to be a valuable additional source of information. Indigenous sources also mention repeatedly the importance of the smallpox epidemic.[36] And, according to Nahuatl sources, Cuitlahuactzin, Cortés's most formidable opponent, died of smallpox. Cortés, however, decided not to report this event in his letters to Charles V.[37] Descriptions of the disease, however, were generally vague because the disease had never been observed and there was not a specific word to describe it.[38]

In the debate over the importance of smallpox, scholars also have turned to recent findings in genetics and public health to better explain this virgin soil epidemic. It has been proven by geneticists that certain pathogens—such as measles and smallpox—tend to hit worst the populations with less genetic diversity. Amerindian populations were unusually homogenous genetically in contrast to Africans or Europeans; this was in great part because they had lived in isolation from the rest of mankind for such a long time.[39] Viral epidemiologists explain that in the case of the "smallpox virus it adapts quickly to a host's immunological response—not mutating into a new strain, but rather preparing for a battle with other hosts of nearly identical genetic makeup."[40] This predisposition was compounded by exposure that was simultaneous and from multiple sources, particularly from members of the same families.

New research in the discipline of public health highlights the importance of social agency in treating an epidemic. An unfortunate factor for Indian populations affected by the first epidemic was that they had no notion of how to care for those who fell ill with this particular disease; with smallpox, nursing (e.g., water; food; and clean, warm clothing) reduces mortality. This meant that entire families became ill and there was no one left to care for the sick, which in any case, they did not know how to do. As survivors became immune and learned how to care for smallpox victims, the Indian populations were better able to cope with disease outbreaks and, not surprisingly, mortality declined substantially in the following outbreaks.[41]

The findings of recent research in genetics and public health thus bolster the analysis of "[e]vidence from a wide variety of sixteenth-century Spanish and Nahuatl sources [pointing] to a single conclusion: the smallpox epidemic of 1520 ranked among the three worst demographic crises of the century and leaving aside the controversy over the size of Amerindian populations at contact, there emerges a broad agreement in the Spanish and Nahuatl narratives and in the patterns of decline sketched by historians."[42] The debate is not over, but recent studies have not defied this conclusion.

Did Smallpox Travel into Peru Faster than the Spaniards?

Traditional historiography asserts that in the case of Peru, smallpox preceded the arrival of Spaniards. It was believed that Huayna Capac, the Inca ruler of Andean regions spanning from modern-day Ecuador to Chile, died of smallpox between 1525 and 1527, while Francisco Pizarro did not land on the coast of Peru until 1532.[43] For the alleged cause of Huayna Capac's death to be true, it would have been necessary that smallpox traveled thousands of miles, spreading through face-to-face contact across regions with low densities of population and different climates and altitudes. Epidemiologists today have proven that this hypothesis is both improbable and impossible.[44] Huayna Capac died of another contagious disease, most likely a lethal mutation of a local pathogen.

Prior to the arrival of Spaniards, the Inca ruler had undertaken major construction projects and had altered agriculture practices in the Andean empire; the magnitude of these changes caused ecological alterations conducive to the development of new strains of already existing diseases.⁴⁵ And indigenous and Spanish sources concur that by the time Pizarro arrived in Peru, the population of the Inca empire was debilitated by years of civil war and disease.⁴⁶

Syphilis: The Legacy of the New World to Mankind

Syphilis was traditionally believed to be a disease imported from the New World into the Old.⁴⁷ Several chronicles dating to the end of the fifteenth century describe this disease, called "the pox" at the time, as one attacking with the spread and rapidity of a plague.⁴⁸ Venereal syphilis, however, is not a plague in the strict sense of the term because it is a sexually transmitted disease. Its rapid spread throughout the Old World since the end of the fifteenth century was caused by social rather than biological circumstances. Venereal syphilis arrived in Europe at a time when wars were being fought in different fronts, and with wars came the relaxation of sexual habits among the different populations.

During the twentieth century, in light of the development of research on epidemiology, it has been found that there are four different strains of syphilis and three of them were already present in Europe, Asia, and Africa.⁴⁹ Since then, some historians began to question the plausibility of venereal syphilis being a New World legacy to mankind. Two points, however, are pertinent to those who question the origins of this disease. The first is that prior to 1492 no description of symptoms similar to those of venereal syphilis had been recorded in Europe. The second is that also prior to the sixteenth century, no signs of *Treponema pallidum* (the organism that causes venereal syphilis) have been found in skeletal remains of any population outside of the Western Hemisphere. In the Americas, *Treponema pallidum* was endemic, but when it got to the Old World it became epidemic, as this population had not been exposed to this pathogen. This conclusion is possible thanks to research done by physical anthropologists specializing in osteopathology. Through the analysis of skeletal remains it has become possible to assess the health of ancient populations and, in this particular case, ascertain the origin of venereal syphilis.

Environmental Changes and Demographic Collapse

Crosby's work also advances the argument that the introduction of new flora and fauna by Europeans into the American continent contributed to the demise of the indigenous population. When Spaniards came to the New World they brought with them the plants and animals with which they were familiar in order to replicate in the Americas the way of life they had in Europe insofar as this was possible. Many of the Spanish common staples, such as wheat, wine, and olives, could not be cultivated in many of the American regions, such as the tropics, so it should come as no surprise that Spanish settlements tended to develop more in temperate areas. Interestingly, Indians were not fond of European food plants and rarely included them in their diet. They eventually cultivated them for tribute purposes. Europeans also brought the livestock they depended on for meat and transportation. Unlike the case of food plants, Amerindians enthusiastically accepted European livestock.[50]

The introduction of new crops and animals in the Americas negatively altered many ecosystems. This does not mean that traditional agricultural practices of Indians were necessarily earth-friendly or self-sustaining in the way we know of them today. They did what they needed to do to survive, and in many cases their survival meant the degradation of their environment. Indeed, land erosion due to agricultural practices was one of the main reasons why for centuries prior to the arrival of Europeans, different Indian populations had migrated throughout the continent: It was an ongoing search for new lands to cultivate. The collapse of many Mayan settlements towards the end of the Classic period (AD 300 to 1000) is perhaps one of the better-studied cases of this phenomenon.[51]

The new interest in the introduction of flora and fauna has inspired studies of the environmental consequences of the encounter of both worlds. One of the best-known works on this topic is Elinor Melville's *A Plague of Sheep*. Her work on the *Valle del Mezquital* in central Mexico presents the quintessential example of environmental degradation in the Western Hemisphere caused by the introduction of livestock during the sixteenth century. Melville argues that the central valley of Mexico experienced a decline in human population at the same time that animal populations—livestock of European origin—were growing.[52] In 1548, the *Valle del Mezquital* was depicted as a "densely populated region where grains were grown in large quantities for subsistence, exchange and tribute.[53] The subsequent introduction of livestock in the region, especially sheep, outstripped the ability of vegetation to regenerate. Further, access to and exploitation of natural resources were altered permanently. The landscape was changed so rapidly by this ungulate eruption that by the last quarter of the sixteenth century it was described as region only suitable for sheep grazing.[54] The frail ecological equilibrium of semi-arid regions was disrupted, and these lands became impoverished.

Experiences similar to that of the *Valle del Mezquital* were repeated throughout different parts of the Americas during the sixteenth century, especially in temperate regions, where grazing animals and European crops thrived, and Spaniards preferred to settle.[55] Indian populations that inhabited these regions at first declined in numbers due to disease and, when they finally recuperated, became populations of impoverished peasants.

Alterations in the ecosystem can influence the way diseases are transmitted and can foster mutations in local pathogens. For example, since the 1970s epidemiological research has found that arenaviral diseases that manifest through hemorrhagic fevers that have affected populations in different regions in the American continent are related to changes in agricultural practices that affected ecosystems.[56] These diseases were not an import from the Old World but rather were different strains of preexisting pathogens found in different regions of the Americas.[57] They spread through rodent reservoirs or other native hosts.[58]

In studying the etiology of hemorrhagic fevers cases in the 1970s, it has occurred to researchers that this is not a twentieth-century phenomena, since this is not the only period in which there have been stark alterations in the ecosystem due to changes in agricultural practices, and not necessarily to the introduction of new pathogens or the emergence of new strains of viruses. Over the past decade, historians and medical doctors in joint collaboration have undertaken the analysis of sixteenth-century disease outbreaks in Mexico to detect the origin and symptomatology of each outbreak and ascertain if these were smallpox, typhus, or hemorrhagic fevers. They have conducted a thorough study of Mexican rodents and mammalian bionomics as well as a revision of anthropological data describing postcontact Aztec agrarian practices, and they have examined published codices that illustrate the symptoms of each outbreak during the sixteenth century.[59] Interestingly, they found that the Nahuatl term to define the disease *cocoliztli* was a broad definition of pest, not a Nahuatl term to define smallpox, measles, or typhus. In conjunction with the Nahuatl records are autopsy reports of Spanish doctors who witnessed the 1576–1580 outbreak, which strongly suggest that it was a disease previously unknown to them, substantiating the hypothesis that it was not smallpox. Another point that favors the notion that the epidemics were not typhus or smallpox is that the "geography of disease is not consistent with the introduction of an Old World virus to Mexico, which should have affected both coastal and highland populations. In both the 1545 and 1576 epidemics the infections were largely absent from the warm, low lying coastal plains on the

Gulf of Mexico and Pacific Coast."[60] This interdisciplinary research group has thus concluded that two of the most deadly outbreaks of the sixteenth century, those of 1545–1548 and 1576–1580, were most likely hemorrhagic fevers and not smallpox or typhus. This form of interdisciplinary research has yet to be applied to study other disease outbreaks in the Western Hemisphere, but such interdisciplinary work promises to shed more light on the existing links between the spread of disease and ecological alterations.

In recent centuries climatic conditions in the world have been altered by human action; the implications of climatic change are severe and are raising our awareness of the need to stop these alterations. There are other cases, however, where changes in climate are natural phenomena, such as the Little Ice Age that occurred from circa 1430 to 1850.[61] Earlier research on the Little Ice Age focused on Northern Europe, where lower-than-average temperatures were correlated with crop failure.[62] Given that climatic history is a relatively new field, it was not until recently that study of the effects of the Little Ice Age in the American continent began. Since the area that was most affected by this phenomenon was the North Atlantic of 50 degrees north latitude, at first it was thought that the Little Ice Age had little effect in Mexico and parts south because they lay below the 50 degree north latitude. However, new research on climate change has shown that there is a correlation between cooling in northern latitudes and drying in tropical latitudes.[63]

A way to assess this correlation is by looking at tree ring evidence that facilitates the reconstruction of levels of precipitation. This research is conducted by dendrochronologists. For the case of Mesoamerica there are pre-Columbian indigenous documents that contain thorough depictions of natural phenomena, which include anomalous climatic events such as storms and droughts.[64] Over the past decade, a joint collaboration between dendrochronologists, epidemiologists, and historians studying the American continent have been able to develop a "continuous, exactly dated tree-ring reconstruction of maize yield variability in central Mexico from 1474 to 2001 that provides a new insight into the history of climate and food availability in the heartland of the Mesoamerican cultural province."[65] The findings of this group indicate that the Little Ice Age affected climate and thus food production and the spread of diseases. Although these changes were not caused by human action, in those places where there were at some point alterations of the landscape by changes in agricultural practices or the introduction of new animal species, the negative effects were probably more exacerbated.

Conclusions

Today we have a clearer picture of the consequences of the arrival of Europeans to the New World. Over the past century, new fields of knowledge have enabled us to understand in more depth how life in the American continent was transformed over the first century of exchanges with the rest of the world. Nevertheless, we have become more aware that more questions remain to be investigated. We still do not have an accurate estimate of the size of the population of the Western Hemisphere at the time of the arrival of Europeans. Nor do we have knowledge of the exact magnitude of the decline in the native population during the first century of the encounter of both worlds. What we can assert is that the near demise of indigenous populations in the Americas cannot be explained by a simple and unique cause; rather, the demographic collapse resulted from a complex set of multiple causes.[66]

The causes are environmental as much as human. Some environmental aspects were a by-product of human action, such as the introduction from Europe of animal species unknown to the Americas and new pathogens, but others were endogenous, such as the effects of the Little Ice Age, the series of climatic variations that were taking place independent of the coming of Europeans but that did exacerbate the implications of European actions for the environment. Diseases were a big killer of Indian populations, but not all epidemics or pathogens were of European origin; some were mutations of indigenous germs, and such mutations were in some cases a response to environmental alterations produced by human factors.

Human causes exacerbated the decline in population independent of the spread of diseases, such as "the weakening of reproduction, expulsion and forced migration of the indigenous people into hiding or into inhospitable areas."[67] All of these causes were associated with the patterns of European conquest and settlement.

Nonhuman and demographic causes of the decline in the population of the Americas that natural and social scientists have proposed do not preclude all aspects of the Black Legend. As McCaa explains, "[D]isease cannot be understood without taking into account the harsh treatment (forced migration, enslavement, abusive labor demands, and exorbitant tribute payments) and ecological devastation that accompanied Spanish colonization." Las Casas's *Short Account* exaggerated the size of the Amerindian population and the number of deaths inflicted by the Spaniards' brutal treatment, but his reports had a basis in reality. Las Casas's Black Legend remains the point of departure to study this passage of the history of the Americas.

The purpose of this essay has been to present the reader with an example of how multidisciplinary tools can be employed in teaching the history of the Americas. In this particular essay on the history of the first encounters, I demonstrate the usefulness of research findings from the natural and social sciences, and their relevance for comprehending the causes of the decline of the native population during the first 100 years after the arrival of Christopher Columbus to the New World. The work of geologists, geographers, epidemiologists, biologists, anthropologists, demographers, linguists, and historians all can contribute to understanding how and why the population of the Americas declined so much during the sixteenth century. It is moreover illuminating to adopt such an integral perspective for all of the Americas, both North America (Mexico) and South America (Peru).

The findings presented here are complementary to traditional history curricula that explain the political, economic, military, and religious aspects of the first century of European presence in the Western Hemisphere. One benefit of this approach to teaching is that it can show students that a broader and more integral perspective of history is relevant for modern life. We are presently facing the possibility of traumatic changes in our ways of life across the Americas as a result of climatic, environmental, and epidemiological developments, which are interrelated with economic, political, and cultural conditions. The challenge is to comprehend these imminent changes from an interdisciplinary perspective that incorporates the work of so many disciplines.

Notes

1. Alfred W. Crosby, Jr., *The Columbian Exchange: Biological and Cultural Consequences of 1492* (Greenwood Press, Westport Connecticut, 1976), Chapter 2.
2. Ibid., chapter 4.
3. Ibid., 15.
4. Ibid., xiv.
5. Ibid.,4–5.
6. James Lockhart and Stewart B. Schwartz, *Early Latin America: A History of Colonial Spanish America and Brazil* (Cambridge: Cambridge University Press, 1983), 32–33.
7. Anthony Pagden, "Introduction," *A Short Account of the Destruction of the Indies*, by De las Casas, Bartolomé, ed. and trans. Nigel Griffi (Penguin Classics, London, 1992), xviii–xxi.
8. Ibid., xxxvi.
9. "At a conservative estimate, the despotic and diabolical behavior of the Christians has, over the last forty years, led to the unjust and totally

unwarranted deaths of more than twelve million souls, women and children among them, and there are grounds for believing my own estimate of more than fifteen million to be nearer the mark" (Las Casas, *Short Account*, 12)

10. Pagden, "Introduction," *A Short Account*, xiii.
11. Tlaxcala Indians gave the name "Motolinía" to Toribio de Benvanente, meaning "poor one" in their language. He adopted it, as he decided to live "entirely devoted as a follower of the Poor Man of Assisi to the conversion of the Aztecs." Borgia Steck, Francis "Introduction" to *History of the Indians of New Spain*, by Motolinía, Toribio de, trans. Francis Borgia Steck (Washington, D.C.: Academy of American Franciscan History, 1951), 6–7.
12. The *History of the Indians of New Spain* was first published in 1858, and the *Memoriales* were first published in 1903; see Robert McCaa, "Spanish and Nahuatl Views on Smallpox and Demographic Catastrophe in Mexico," *Journal of Interdisciplinary History* 25.3 (Winter, 1995): 415; Massimo Livi-Bacci, "The Depopulation of Hispanic America after the Conquest," *Population and Development Review* 32.2 (June 2006): 203.
13. See López de Gómara, *Historia de las Conquistas de Hernando Cortés* (Mexico: Impresta de la testamentaria de Ontiveros, 1826); and Díaz del Castillo's, *True History of the Conquest of New Spain* (Berlin: Asher & Co, 1908).
14. See Georges Baudot, "Introduction," *Historia de los Indios de la Nueva España*, by Fray Toribio de Motolinia (Madrid: Clásicos Castalia, 1985), 40; McCaa, "Smallpox ," 415.
15. Las Casas, *Short Account*, 3; Motolinía, *Historia de los Indios*, 35.
16. Pagden, "Introduction," *A Short Account*, xiii,; David A. Brading, *The First America: The Spanish Monarchy, Creole Patriots and the Liberal State, 1492–1867* (Cambridge: Cambridge University Press, 1993), 458, 592, 611.
17. John S. Marr and James B. Kiracofe, "Was the Huey *Cocoliztli* a Haemorragic Fever?" *Medical History* 44 (2000): 342; for a summary of epidemics occurring in sixteenth-century Mexico during the first century of the colonial period, see Charles Gibson, *The Aztecs Under Spanish Rule* (Stanford: Stanford University Press, 1964), 448.
18. Robert McCaa, "The Peopling of Mexico from Origins to Revolution," in *The Population History of North America*, eds. Haines, Michael and Richard Steckel (Cambridge: Cambridge University Press, 2000), 252; Livi-Bacci, "Depopulation," 199.
19. Sherburne F. Cook and Woodrow Borah, *Essays in Population History: Mexico and the Caribbean*, vol. 1 (Berkley: University of California Press, 1971), 1.
20. See *Essays in Population History*, vols.1–3 among the most widely known works.
21. Cook and Borah, *Essays*, vol. 1, 17.
22. Ibid., 77.

23. Cook and Borah, *Essays*, vols. 1–3.
24. This classification was minted by McCaa in "The Peopling of Mexico."
25. McCaa, "Peopling," 252.
26. Ibid., 254.
27. On Yucatán, see Márquez and Del Angel " Height among Prehispanic Mayas of the Yucatan Peninsula"; on Teotihuacán, see Storey, *Life and Death in the Ancient City of Teotihuacán: A Modern Paleodemographic Synthesis* (Tuscaloosa: University of Alabama Press, 1992).
28. McCaa, *Peopling*, 244.
29. Woodrow Wilson Borah, "Introduction," *Secret Judgements: Old World Disease in Colonial Spanish America*, eds. Noble David Cook and W. Gorge Lovell (Normon: University of Oklahoma Press, 1991), 9.
30. Ibid., 8.
31. Crosby, *Columbian*, 37.
32. Francis J. Brooks, "Revising the Conquest of Mexico: Smallpox, Sources, and Populations," *Journal of Interdisciplinary History* 24.1 (Summer, 1993): 2, 4.
33. Ibid., 21–22.
34. Ibid., 29.
35. See James Lockhart, *The Nahuas after the Conquest* (Stanford: Stanford University Press, 2000).
36. McCaa, "Smallpox," 400–401.
37. Ibid., 407.
38. Ibid., 408.
39. Crosby, *Columbian*, 37; McCaa, "Smallpox," 420.
40. McCaa, "Smallpox," 419.
41. Crosby, *Columbian*, 53; McCaa, "Smallpox," 421.
42. McCaa, 'Smallpox," 428–429.
43. Crosby, *Columbian*, 51; Livi Bacci, "Depopulation," 208.
44. Livi-Bacci, "Depopulation," 215.
45. Marr and Kiracofe, "Huey Cocoliztli," 362.
46. Livi Bacci, "Depopulation," 226.
47. Crosby, *Columbian*, 123.
48. Ibid., 149.
49. Ibid., 141.
50. Ibid., 74.
51. Rebecca Storey, Lourdes Márquez Morfin, and Vernon Smith, "Social Studies and the Maya Civilization of Mesoamerica: A Study of Health and Economy of the Last Thousand Years," in *The Backbone of History*, eds. Steckel, Richard H. and Jerome C. Rose (Cambridge: Cambridge University Press, 2003), 286–287; Crosby, *Columbian*, 111.
52. Elinor G. K. Melville, *Plague of Sheep: Environmental Consequences of the Conquest of Mexico* (Cambridge: Cambridge University Press, 1994), 6.
53. Ibid., 158.
54. Ungulate eruptions refer to how grazing animals move into new ecosystems. Melville, *Plague of Sheep*, 39, 159.

55. Melville, *Plague*, 2; Crosby, *Columbian*, 77, 110.
56. Marr and Kiracofe, "Huey Cocoliztli," 343.
57. Ibid., 355.
58. Marr and Kiracofe, "Huey Cocoliztli," 354; Rodolfo Acuña-Soto, David W. Stahle, Malcolm K. Cleaveland, and Matthew D. Therrell, "Megadrought and Megadeath in 16th Century Mexico," *Emerging Infectious Diseases* 8.4 (April 2002): 361.
59. Marr and Kiracofe, "Huey Cocoliztli," 343.
60. Ibid., 349.
61. Susan L. Swan, "Mexico in the Little Ice Age," *Journal of Interdisciplinary History* 11. 4 (Spring, 1981): 633.
62. Ibid., 638.
63. Ibid., 639.
64. Therrell, Stahle, and Acuña-Soto, "Aztec Drought and the 'Curse of One Rabbit,'" *American Meteorological Society* (September 2004): 1263.
65. Matthew D. Therrell, David W. Stahle, José Villanueva Díaz, Eladio H. Cornejo Oviedo, and Malcolm K. Cleaveland, "Tree Ring Reconstructed Maize Yield in Central Mexico: 1474–2001," *Climatic Change* 74 (2006): 493.
66. Livi-Bacci, "Depopulation," 224.
67. Ibid., 226.

PART 3

Programs and Pedagogy

CHAPTER 9

Beyond National Borders: Researching and Teaching Jovita González

Heather Miner and Robin Sager

This co-written essay provides a multidisciplinary introduction and biographical context for those interested in reading a recently rediscovered author of the U.S.-Mexican borderlands, and for teachers who might want to add a hemispheric Americas author to their syllabi.[1] Jovita González was a Mexican American woman from rural Texas who worked in the mid-twentieth century as an anthropologist, historian, folklorist, and novelist. She is an increasingly pertinent figure to study and teach, due, in part, to her unique professional and intellectual journey. In addition, the diversity of academic fields represented in González's scholarly and fictional works makes her an apt figure to teach in a range of undergraduate and graduate classrooms. Our goal is to offer avenues for studying González within classroom conversations about national foundational myths and histories, as well as discussions about the development and practice of academic disciplines. Thus, this essay provides a cultural and biographical introduction to González that is necessary in order to understand this author within her Mexican American background and the academic context in which she worked, and to call attention to how these contexts shaped her works, and, in particular, the historical novel *Caballero*. We focus on the thematic attention González pays throughout her literary, folklore, and historical works to the expansion and transformation of the U.S. nation to argue that González's writings pose a challenge to the master narrative of historical inquiry and nation building in the twentieth century.

The modern critical reception of González was spearheaded by José Limón and María Cotera, scholars who have been instrumental in recovering archival texts and initiating attempts to understand the personal history of a figure whose work and life story are fascinating but also somewhat obscure. Limón and Cotera pioneered attention to González with their collaboratively produced edition of *Caballero: A Historical Novel*, a work that remained unpublished during González's lifetime.[2] In addition to disseminating previously unknown texts and organizing archival holdings, Limón and Cotera have each initiated distinctive approaches to studying González's life and work. Limón has drawn on his extensive knowledge of González's life to produce numerous interpretational essays highlighting the importance of González's writings on Mexican American folklore. Cotera has been instrumental in examining the complex marginization of González in mainstream and feminist intellectual circles.[3] In addition to highlighting González's importance in gender and folklore studies, Leticia Magda Garza-Falcón is a notable early scholar of González; Garza-Falcón has focused on González's political importance to argue that González helped to create Chicano studies because González's life "offer[s] contradictions to the restrictive, flattened pictures of Mexicanos that have for the most part prevailed in U.S. history and narrative fiction."[4] These early and significant advocates of González's work offer key interpretations of her disparate writings and suggest many avenues for further engagement, a project already being undertaken in a range of academic conversations. Scholarly articles on González appear in Chicano, modernist literature, American history, and ethnic studies journals, a noteworthy range.[5] As a result of this scholarship on González's work, her texts are increasingly known to an academic audience but are as yet seldom taught in the classroom. Our attention in this essay to the transnational and multidisciplinary aspects of González's work grows out of our awareness of a gap in recent work on González, which either focuses on her novels or her nonfiction, anthropological writing, but overlooks the important interconnections between these elements of her work. This essay is intended to help bridge this divide and therefore provide a guide for exploring previously untapped avenues for teaching González in the classroom.

In order to understand the tensions, both disciplinary and geographic, within González's *Caballero*, as well as her other works, it is necessary to view González within the biographical and cultural background that furthered the development of these themes. González was born at the beginning of a century marked by struggles over the definition and management of populations on the local, national, and global levels: 1904, the year of her birth, also witnessed the beginning of construction on the Panama Canal, the establishment of a chair in the study of eugenics at the University College London, and the opening of the Louisiana Purchase Exposition (also known as the St. Louis World's Fair), complete with a "human zoo" displaying Native Americans and pygmies from the Philippines.[6] González experienced the impact of transnational spatial and racial encounters of these types on a local level in her hometown of Roma, Texas, as the remapping and perception of global communities was felt in a place seemingly divorced from larger historical developments. Roma was a ranching community with a population of 521, situated on the Texas-Mexico border, a setting that provided González with an opportunity to observe the struggles over land between Mexican and Anglo-American groups, as well as the political neglect of immigrants.[7] Although González and her family moved to San Antonio in 1910, the scenery and peoples of the South Texas borderlands were a fixture in González's work throughout her life.

González's formative years in Roma and San Antonio demonstrate that she continually struggled to reconcile her sense of class identification with the realities of her social environment. Though Gonzalez was a fifth-generation descendant of a family granted land by the Spanish crown and, along with other family members, believed that she came from aristocratic Spanish lineage, González was born into a struggling, lower-middle-class family.[8] Her family's belief in their august heritage could not protect González from the reality that Mexican Americans were often treated as second-class citizens in Texas and throughout the American South. As historian David Montejano argues, education officials in early-twentieth-century Texas failed to meet the needs of the Mexican American community and instead funneled students into schools with fewer resources, shorter terms, and inadequate compulsory education.[9] Yet while González would have certainly experienced these discriminatory challenges in her primary and secondary education, her later writings demonstrate that she was not an advocate for equal opportunity either. As an adult, González clung to a Spanish heritage that sustained her sense of class superiority. Her nascent elitism, held from early childhood onwards, narrowed her ability to empathize with the experiences of the Mexican and American working classes, and resulted in biased accounts of these groups within her writings, as discussed later in this essay.

After completing her secondary education, González decided to pursue a bachelor's degree at the University of Texas at Austin (UT). González entered UT in 1921 with the intent to focus on literature within the Spanish Department under the direction of Lilia Casis, an established professor of advanced linguistics and classical Spanish literature.[10] After only a short period in the program, González was forced to take a leave of absence from the program due to financial circumstances. During this gap in her studies, González taught at local San Antonio and Austin schools, including Our Lady of the Lake College, where she also went on to earn her B.A. González maintained contact with Casis during this period, even enrolling in summer school at UT. However, when González finally returned to UT full-time in the late 1920s to pursue her M.A., she was no longer working within the Spanish Department and instead chose to pursue a degree in history. Her decision to change fields was no doubt influenced by her growing professional relationship with Texas folk scholar J. Frank Dobie, whom she met in 1925.[11]

Though nominally linked to the English Department at the University of Texas, Dobie was an important figure in promoting the Southwest as a natural site for ethnographic inquiry within the emergent discipline of anthropology in the United States. Working on his own M.A. in 1913 at Columbia University, Dobie was introduced to the practices of cultural anthropology he would later promote in his own work and the work of his students.[12] Dobie's brief time at Columbia overlapped with the tenure of Franz Boaz, the "father" of modern anthropology who famously developed the methodology of the young discipline.[13] As with other early anthropologists, Dobie became increasingly interested in accentuating the importance of local histories within the study of a nation's peoples in order to preserve elements of cultures that were in constant flux. This attention to indigenous histories and cultures was reflected in the creation of amateur local history societies across the United States in the early twentieth century and, in Texas, with the founding of the Texas Folklore Society (TFS) in 1909.[14] Upon his return to Texas in 1914, Dobie joined the society, the promotion and professionalization of which would become a large of part of his, and Jovita González's, life's works.[15]

While working under the direction of Dobie at UT in pursuit of her master's in history, González became interested in recording the Southwest border culture in which she had spent her childhood. As she would write in a short autobiography later in life, prior to working with Dobie, "the legends and stories of the border were interesting so I thought just to me. However, he made me see their importance and encouraged me to write them."[16] González published much of this early work in the *Southwest Review* and in TFS publications (a large part of González's writings in these journals are committed to recording the society of lower- and working-class Mexican Americans, a somewhat ironic dedication, given her own conservative classism). The TFS, in tandem with publishing González's work, mandated strict guidelines on how an untrained observer was to properly conduct ethnographic research. A 1911 bulletin provided rules that changed little over the society's lifetime, instructing that, "For the collector of Folk-Lore, the most important virtue is accuracy; and the value of any contribution is destroyed if it is not given just as it was told or sung or described, with no change whatever, even when such change seems necessary to make sense. Second to accuracy, it is of great importance that full information be supplied... a transcript in the exact words of the informant is best—colloquialisms, meaningless words, mistakes, and all—and, in the case of ballads and much of the other work, such exactness is necessary."[17] These rules influenced González's writings throughout her lifetime, and the early sketches she published accorded with the society's desire to preserve the "body of tradition which is handed down by word of mouth from generation to generation... the myths, legends, popular beliefs, folk-songs, and folk-tales of all countries."[18] González became an active member of the TFS, serving as vice-president and then president of the society, from 1930–1932. She was the first female president, and one of a few Mexican American members in the organization. Though this accomplishment is noteworthy, it is clear that hers was mainly a titular position; the TFS was directed and managed almost entirely by Dobie himself.[19]

While González was still a master's student at UT and becoming increasingly active in the TFS, she was also growing gradually more frustrated by the ethnographies of Mexican Americans being presented by the society. González felt that border life demanded compassion by the author to be truthfully represented; the fragmentary documentation of testimonials published by the TFS overlooked the larger historical narratives that were influencing the tales and changing how they were told, both by the informant and the writer.[20] It was with this mindset that González approached the research and writing of her master's thesis. In 1929 she traveled along the impoverished and rapidly transforming parts of the Mexican-American border to begin the research that would ultimately provide the material for her M.A. thesis, *Social Life in Webb, Starr and Zapata Counties*. Despite the sociological title, the work is an odd mixture of ethnography, history, and personal analysis. In six chapters, González provides a history of the region as well as a description of the often-violent encounters between Mexican peoples of the border and Anglos. The thesis, according to González, was a result of studying the indigenous culture on the border and "a lifetime of love and understanding for my people, the border people."[21] The generic hybridity of this work illustrates González's desire to push beyond the disciplinary limitations her academic training had attempted to impose. Writing in the late 1920s and 1930s, González's work considers if one can most accurately represent one's culture as if an outsider, for whom everything is unknown and thus ostensibly viewed through the objective lens of the anthropologist, or from the distinctive identity of an insider who is capable of contextualizing specific details as a cultural historian.

González's inability to reconcile these issues within her work was only complicated by the fact that she had to switch advisors during the thesis process. Perhaps because of Dobie's tense relationship with the UT administration (having been initially barred promotion because of his lack of a doctoral degree in 1925, Dobie returned to UT in 1929 with token support from the university faculty[22]), González completed her thesis with Eugene C. Barker as her advisor. Barker was a noted Texas historian who was active in the creation of the American Historical Association and the managing editor of the *Southwestern Historical Quarterly*; though his work, by the standards of modern historians, often seems slanted and subjective, his publications were considered leading examples of accurate historical scholarship in the 1930s.[23] González's thesis, from Barker's perspective, represented a failed attempt at historical scholarship, as it "did not have enough historical references."[24] Barker was uncertain about the thesis, in part, because the work, in a similar fashion to its author, displayed obvious interdisciplinary tendencies at a time when disciplinary rigidity was the rule. Thus, the thesis only passed after the intervention of another faculty member and González's longtime friend Carlos E. Castañeda, who suggested that the thesis "will be used in years to come as source material."[25] Barker did not concur, only begrudgingly agreeing that it was "[a]n interesting but somewhat odd piece of work."[26] However, in the end, Barker approved the thesis, and in 1930 González received her M.A.

Barker's negative reaction was not representative of the attention that González's early work garnered elsewhere. As part of her professional progression in academia, from her TFS presentations to her master's thesis, González actively cultivated her claims to authenticity and insider status. As Cotera notes, "González's 'authenticity' as a daughter of ranchero culture constituted the very foundation of her ethnographic authority, and she was not beyond capitalizing on this patina of authenticity to further her own position."[27] In particular, González's desire to interpret, not simply preserve, the cultural products of border life, and blending of subjective analysis, history, and ethnography coincided with the evolving goals of the Rockefeller Foundation. Upon the recommendation of Dobie, and following an interview with David H. Stevens, the director for the foundation's Division of the Humanities, González received a grant in 1934 to expand upon her thesis research and continue her study of life in South Texas.[28] This grant was a part of the foundation's larger redistribution of funds away from traditional humanistic research (which Stevens claimed "buttress[ed] scholasticism and antiquarianism in our universities"[29]) and into the exploration of modern daily life around the globe.[30]

The result of González's grant was *Dew on the Thorn*, a text that she worked on throughout the 1930s but never published in her life. *Dew on the Thorn* is akin to González's thesis in its attempt to incorporate historical and ethnographic details, but strikingly diverges with its aesthetic perspective. The text is a collection of folk tales, myths, and anecdotes about border life, all loosely linked through a fictionalized account of the Olivares family. In this heterogeneous text, the fictional plot follows two star-crossed lovers whose romance becomes the pretext for an exploration of ranchero life. The autobiographical elements of the work are patent: It opens in the year of González's birth, 1904, and is concerned with the troubles faced by a traditional aristocratic Spanish family upon the incursion of Anglo-Americans into the border space, a theme drawn directly from González's own familial background. In addition to the blend of history and fiction in the tale, though, *Dew on the Thorn* diverges from González's earlier work in its deliberate political and feminist overtones. The emergent leader of ranchero life and culture, which is seemingly being torn apart by American democracy and Texas state politics, is Dona Margarita, the matriarch of the Olivares family. In her concluding declaration, González's heroine adamantly states her opinion on the interference of American culture and the increasing racialization of Mexican Americans: "The *Americanos* may come. They may take the land, but our spirit, the spirit of the conquerors, will live forever. Texas is ours. We stay!" (*Dew*, p. 179). Dona Margarita's pronouncement is an echo of that made by González's great-grandmother, on whom the fictional matriarch is based. As González dramatically recounts in a short autobiography, on her deathbed González's great-grandmother Mamá Ramoncita declared to her grandchildren that "'Perhaps they will tell you to go to Mexico where you belong. Don't listen to them. Texas is ours. Texas is our home. Always remember these words: Texas is ours, Texas is our home.' I have always remembered the words and I have always felt at home in Texas."[31] González was forced constantly to negotiate the assertion of her Texas belonging within a nation that increasingly sought to reject Mexican Americans a right of place. Although the text itself mirrors her thesis very closely, the plot of *Dew on the Thorn* is of critical importance because it represents a transition from her desire to simply provide information for cultural education and instead adopts a more aggressive tone reflecting González's developing certainty that Mexican and Anglo-Americans could only live peacefully together if both sides made concerted efforts toward mutual accommodation and understanding, an effort that was often clearly lacking. This belief would later reach maturation within *Caballero* in which González describes not only the path to cultural coexistence, but also the tragedies that can occur due to continual misunderstandings.

The Rockefeller grant and the opportunities that it afforded considerably affected González's professional and intellectual journey. However, during the mid-1930s she also experienced profound changes within her personal life. In 1935, she married Edmundo E. Mireles, a San Antonio educator and UT graduate. Although she continued to publish with the Austin-based TFS, González and Mireles eventually relocated to Del Rio and then Corpus Christi, where they both secured full-time teaching positions in local high schools. In a similar fashion to her intellectual productions, González's approach in the classroom was also infused with interdisciplinarity and she at various times taught Spanish, Southwest history, and English literature courses. In the twenty-one years that she spent as a Spanish teacher at Corpus Christi's W. B. Ray and Miller high schools, González was known not only for instilling a love of the Spanish language within her students, but also for incorporating "Mexico and Spanish culture, history, and folklore" into her lesson plans.[32] With the passage of time and her teaching commitments, it appears as if González's relationship with Dobie and the TFS gradually weakened. By 1937, only five years after her tenure as president, González no longer paid dues to the TFS.[33] Yet González continued to write throughout her lifetime. When she died in 1983, González had few remaining ties to Texas folklorists, apart from her own unpublished writings. Along with her ongoing work on borderland folklore, the project that would occupy González for several decades (at least throughout the 1930s and 1940s, and perhaps longer) was a collaborative tale written with Margaret Eimer (pseudonym Eve Raleigh), *Caballero: A Historical Novel*.[34]

As we have shown in the preceding pages, González's biography illustrates that she is intriguingly distinctive in her development from a schoolteacher in South Texas into a scholar who transgressed traditional intellectual and geographical boundaries. *Caballero* is a novel uniquely focused on Mexican and American relations in the mid-nineteenth century, and is, in particular, a worthy addition to classroom syllabi. *Caballero* adds new possibilities for the teaching of American literature and history from the geographic perspective of the Southwest. The novel portrays the establishment of the U.S. nation, not from Plymouth Rock or Jamestown, but from the violence along the Rio Grande and the politics of Mexican and American integration in Texas.

Caballero was a project that occupied González and Eimer, an Anglo-American woman living in Arkansas, for more than thirty years, much of it done in correspondence between the two women. In 1939, González wrote that the source material for the book "has taken me twelve years to compile from memoirs[,] traditions and of course historical sources into which I have delved at the... University of Texas."[35] Late that same year, González and Eimer signed a contract agreeing to "equally divide all revenues from the novel" should it be published.[36] Unfortunately, though González and Eimer submitted the novel to publishers throughout the 1930s and 1940s, it was continually rejected. One New York literary agent explained the constant rejections to González by stating "A very long novel like this one, must possess a certain great virtue. It must be a literary achievement, which this is NOT."[37] However, in one of her letters to González, Eimer hinted that the rejections might have stemmed not only from the work's unusual length, but also from "a craze for so-called realism" that made a historical romance such as *Caballero* unattractive to publishers.[38]

Although the work may not have been timely, *Caballero*'s trajectory and themes are consistent with González's earlier writings; indeed, the romantic arc of the novel and interest in narrating an alternate history of the expansion of the United States is analogous to the structure of *Dew on the Thorn*.[39] Yet *Caballero* is a much more cohesive novel. *Caballero* is set along the border during the pre-Civil War conflicts between the United States and Mexico from 1846 to 1848, and the novel's historical attention to the annexation of Texas by the United States is narrated through the political and romantic troubles of the Mendoza y Soría family. The family patriarch, Don Santiago, is a traditionally minded autocratic landowner whose attitude towards his wife, family, and dependent peons has changed little from that of his Spanish conquistador ancestors. Don Santiago's politics are entrenched in the text by his constantly repeated belief that "All this that I can see, and far beyond, is mine and only mine" (*Caballero*, p. 33). As Don Santiago surveys his familial ranchero, his constant companion is the abstract representative of Power made literal, "a figure that touched him, and pointed and whispered. Those dots on the plain, cattle, sheep, horses, were his to kill or let live. The peons, down there, were his to discipline at any time with the lash...His wife, his sister, sons, and daughters bowed to his wishes and came or went as he decreed" (*Caballero*, p. 33). Upon temporarily moving his family from the ranchero to the town of Matamoros, Don Santiago and his eldest son, Alvaro, become increasingly drawn into an insurgent war against the occupying U.S. army at the same time as his two daughters, Susanita and Angela, become romantically involved with white Anglo-Americans from the military.

As the conflicts within the family mount, *Caballero* explores the unsettling impact the encroachment of American culture has had on traditional gender and class relations within the closed Mexican community. As the peon Estéban explains, "I heard many things when I was in Matamoros. The *gringos* hire Mexicans, pay them well, and they can do what they please with the money. Not like our masters who charge us for rags of clothes and tobacco at so high a price that we are always in debt to them and we are never paid a wage. Not all the *gringos* hate us, and soon their law says we can vote and our vote as good as the masters" (*Caballero*, p. 220). The critique of the peon system is heightened by the overt suggestion of sexual abuse threaded throughout the novel's portrayal of ranchero life. Following Estéban's protocapitalist speech, another peon looks at a woman nearby and remarks that, "Her master used her when she was little more than a child, and she is still in his command and that of his sons also...the *peon* belongs to the master and none dare say no to him, though he take their daughters and their wives. I for one will watch my chance to work for an *Americano*. Our souls are not our own" (*Caballero*, p. 220). However, the peons soon discover that the racial hierarchy on the ranchero, initially eased as they gained status as paid workers in the United States, is reestablished in mutated ways: though employed rather than bound, the peons' essential daily reality remains unchanged.[40]

The novel makes it a priority to investigate the escalating racialization of Mexicans by the U.S. intruders. Early in the novel, the narrator notes that "Samuel Walker, a brave officer and beloved, was killed in battle in Mexico, and the Rangers wept in their beards when they buried him and vowed further revenge on 'the dirty greasers'" (*Caballero*, p. 78). While indicative of the historically accurate violence directed against Mexican Americans by the Texas Rangers, an Anglo military police force, elsewhere, the novel evidences a more complex relationship to the new racial hierarchies in the state.[41] In addition to the Rangers' illegal forays into Mexico, the novel includes a short episode of civil incursion into Don Santiago's ranchero by "the land-greedy" Americans. After shooting at one man for trespassing and demanding that they leave his land, an American woman "pointed at him and screamed, 'That's what they say, and even if it's theirs they're only Mexicans. We be white folk and this is the United States, ain't it? We got the right of it" (*Caballero*, pp. 194–195). The narrator notes that this "trash" was part of "the land-greedy who justified their rapaciousness with the word "pioneer'. . . squeezed out of a community that refused to support them any longer; the wanderer, fleeing from nothing but himself; the adventurer, his conscious and his scruples long dead. All these, and more, came to Texas like buzzards to a feast" (*Caballero*, p. 195). The complex class relations in the novel are readily apparent in this scene: While the narrator suggests that U.S. domestic politics are the source of racial animosity in the new state of Texas, the novel also sympathizes with Don Santiago's "contempt" for the "land-seeking invaders" (*Caballero*, p. 195). Ultimately, the novel seems to occlude further conflicts in the expanded U.S. nation through a series of upper-class marriages that conclude the novel, but it also issues a final prediction: that the "prejudice, the long-lived, driving new nails in the barrier of racial and religious difference[,] Revenge, and Hatred, and Murder, and Greed" will haunt the newly reformed nation (*Caballero*, p. 331).

González and Eimer's framing appendages to the novel posit a specific audience of Anglo-American readers, unfamiliar both with the Spanish language and traditional aspects of Mexican life. The text opens with two indices: a cast of characters and a glossary of Spanish terms used in the text, including straightforward definitions ("baile. *Dance*"), descriptions of the environment in the novel ("tipichil. *A hard floor of earth. Method: The earth was dug up to at least a foot in depth...*"), or explanatory comments of colloquial phrases ("con diez mil demonios. *With ten thousand devils; a favorite expression synonymous with speed*") (*Caballero*, pp. xxxi, xxxiii, xxxiv). In addition, preceding the foreword that formally opens the narrative is a simple "Sketch of the floor plan of the House of Mendoza y Soría." These introductory entrances into the novel are intended to help the reader understand not just the world of the novel, but, more importantly, the space of the novel. In a text largely concerned with national conflicts over territorial space, the glossary and sketch direct the reader to the most important spaces in the novel, not the sweeping landscapes prized by Don Santiago, but the familial spaces valued by the peons and the Mendoza y Soría women.[42]

Yet the scope of the novel reaches beyond educating the uninitiated Anglo-American reader. The novel attempts to speak to a cross-national audience through its cogent criticisms of both Anglo-American and Mexican communities. González believed that the novel "is the only book of its kind, the Mexican side of the war of 1848 has never been given. We are not partial. We picture the Mexican hidalgo with their faults as well as their virtues, with their racial and religious pride, their love of tradition and of the land which they inherited from their ancestors. We also picture the American officer, their kindness to the conquered race, but we also picture the vandals who followed...the army, hating anything and everything that was Catholic and Mexican."[43] Though the novel's outward construction seems to be limited to Anglo-Americans, the plot is much more ambitious. Partially due to its attention to cross-cultural and supranational tensions, González imagined even greater things for the text: "when it sells, for I feel it will, it will make an excellent moving picture" (writing in 1939, perhaps González envisioned a Texan version of the trajectory Margaret Mitchell's *Gone with the Wind* had taken).[44] It is also possible that González was encouraged by the literary and film success of another Texas novelist, Dorothy Scarborough. *The Wind*, Scarborough's 1925 work about a troubled heroine in drought-stricken West Texas, achieved critical acclaim and was made into a silent movie starring Lillian Gish.[45]

However, it cannot be overlooked that González eschewed the subject of numerous popular films, the Texas Revolution, in order to focus on the Mexican War. Although *Caballero*'s foreword recounts the 1748 arrival of the Spaniards in the space that would later become Texas, the majority of the novel takes place during the U.S.-Mexican War (1846 to 1848). The novel opens by describing a content Don Santiago surveying his vast land holdings in early 1846 and concludes with his death shortly after the signing of the Treaty of Guadalupe Hidalgo on February 2, 1848.[46] By choosing to situate the work within this period, González consciously rejects the Texas Revolution and the resulting Texas Republic periods as the center points of Texas history. Instead, the text argues for an alternative history, focusing not only on Anglo-American achievements, but also on native experiences of history. Academically trained in the 1920s, González was no doubt aware of the master historical narrative arguments made in academic circles, particularly by Eugene Barker. Barker influentially argued that the Anglo-American victories of the Texas Republic period overshadowed the developments of 1846 onward.[47] *Caballero* and its characters proceed as if the republic never occurred, which can be interpreted as González's attempt to claim the Mexican-American War period as one in which historians could, and should, describe the Mexican American point of view.

The temporal and political conflicts that are variously ignored or exaggerated in the novel's historiography are resolved through the multiple marriages in the conclusion: Susanita, the novel's heroine, marries Robert Davis Warrener, a officer from Virginia; her sister Angela, a devoted Catholic, marries Alfred "Red" McLane, an aspiring politician in the newly minted state of Texas; Inez Sánchez, a friend of Susanita's, marries the evocatively named provocateur Johnny White; and Don Santiago's middle-aged, widowed sister Dona Dolores weds a neighboring landowner and old flame, Gabriel del Lago. The superfluity of marriage plots that take over the novel's conclusion is a jarring break from the historical detail that accounts for much of the plot, though the narrator shows some cynicism regarding the romantic overabundance. Following Susanita and Warrener's marriage, the narrator notes, "Susanita's feelings had been an ecstasy of purest love...Angela's was an ecstasy of the soul, for God. Warrener's marriage, to him, was love that was an end in itself, all things bearing upon and bending to it. McLane's was a link to power, a staff of respectability, a means to an end—" (*Caballero*, p. 317). Following in the footsteps of Doris Sommer, who argues that Latin American novels written in the nineteenth and early twentieth centuries linked erotic desire with a message of political desire in romances of national foundations,[48] numerous critics have focused on the marriage of Susanita and Robert in their analysis of the text. In many ways, the novel promotes this focus. The marriage of the beautiful Mexican "white lily" to the slave-owning Virginian, hailing from a plantation with "fine horses and ordered living and entertainments," and who produce "a baby girl with her blond hair" before the end, is figured as the ideal result of cultural integration (*Caballero*, pp. 107, 334). As Vincent Pérez aptly argues, "Interethnic romance in the novel thus projects Mexican-American inclusion within a liberal 'New South(west),'" an imagined biracial society in which the 'premodern' agrarian culture of its Mexican (i.e., southern) minority has not been lost but has served instead to enrich and transform a modern Anglo-American society driven by individualism and capitalism."[49] The fictional blending signifies the assimilation required to form the only recently reimagined community of the U.S. nation.

Yet it is important to remember that this union is only one of many. The range of marriages in the end reflects the multifaceted debates concerning how Texas integration into the United States would be managed by local Mexican communities. While Susanita and Robert represent a foundational fiction, the contractual marriage of Angela and Red is a stark quid pro quo arrangement, with Angela gaining autonomy and money to pursue her appropriately feminine charitable work, and Red using his new upper-class Mexican wife to garner political support. Montejano argues that "[f]or individual families of the Mexican elite, intermarriage was a convenient way of containing the effects of Anglo military victory on their status, authority, and class position. For the ambitious Anglo merchant and soldier with little capital, it was an easy way of acquiring land."[50] While this type of marriage is the one most often found within the historical record (although historians continue to debate the actual degree of assimilation that took place through marital alliances), the novel posits one more model. The marriage of Dona Dolores and Gabriel del Lago is consistently overlooked by critics because of its tacked-on feeling, but it, too, has historical significance. The married older couple, who retire to a rural ranchero, essentially untainted by the political strife around them, become figures preserving the relic of ranchero life. As Armando Alonzo argues, "[t]radition-minded Tejanos preferred marriage as a way to maintain order and stability in their society" in the political upheaval of nineteenth-century Texas.[51] Yet Dolores and Gabriel are notably distant and spatially separated away from the realities of modern life; their marriage is a memorial to a vanishing culture.

Caballero, as with González's previous works, cogently problematizes the too-easy boundaries drawn in interpreting national histories. John González argues that "the discourse of gender critique in [*Caballero*] accomplishes the task of modernizing a specifically class-marked...subjectivity such that national integration becomes possible through the desiring intersubjectivity of the historical romance."[52] However, the multifaceted challenges to an array of nationally specific groups in the novel suggest a more cogent critique of national integration. From the U.S. army's illegal incursions into a Catholic nunnery in Mexico City, where traditional patriarchs attempt to lock away rebellious daughters, to the uncontrolled violence of the legally sanctioned Texas Rangers, *Caballero* consistently shows the power of national institutions only to underscore the violence and corruption endemic to them. This is not a novel intent on buttressing the identity of the young U.S. nation, nor Mexican identity either. *Caballero* rejects the model of community embodied in national institutions and aristocratic traditions, refusing to define the heterogeneous community in Texas through a falsely organic connection to a cohesive past or shared qualities and values. *Caballero*, and indeed much of González's work, though intent on presenting the Mexican American perspective, ultimately decenters the nation as a source of communal and cultural identity. In *Caballero*,

Mexico is a place of almost feudal organization that is isolated and stratified, while the United States' sense of manifest destiny results in a prurient form of nationalism. Both nations are ultimately rejected. González's work deconstructs the idea of the nation as a whole, unified entity with specific borders and cultural signifiers.

Rather, González advocates for the representation of cultural identities rooted in local traditions that connect local practices with larger cultural identities. Warrener's interest in translating Michael William Balfe's *The Bohemian Girl*, an opera first performed in London (and, interestingly, originally based on a Miguel de Cervantes's novel *La Gitanilla*), from English into Spanish appears as a distraction from the main narrative. Yet this act is a significant part of representing the social and aesthetic significance of artistic dialogue. This moment of cultural exchange, and the shared appreciation of Anglo and Spanish histories, is suggestive of González's relationship to her readers. The folklore told on the Olivares ranch in *Dew on the Thorn*, or the intricate Christmas traditions described in *Caballero*, are imbued with cultural importance beyond the local through the value emphatically placed on such moments of seemingly unimportant details. In the moments when the violence of history is quieted, cultural appreciation, the root of true integration, takes place. In emphasizing the role of the local, and even the specific space of the home sketched into the text, González suggests that integrating disparate cultures cannot be accomplished through the violent efforts of hegemonic institutions. Ultimately, cultural integration can only occur through valuing and attempting cultural understanding.

As we have shown throughout this essay, González's usefulness to the classroom lies both in the example of her unique intellectual journey as well as in her ability to problematize disciplinary and national boundaries and borders in her writings. We have emphasized the broad value of González's work because we believe that she offers something rarely available to U.S. students: an alternative history of the foundations of North American national histories from a Mexican American perspective. *Caballero*, a product of González's life experiences and historical acumen, rewrites the hegemonic understanding of Texas and American history in a way that does not celebrate the victories of one group or another, but instead relishes the possibility of true cultural integration. The representation of this history, and the acknowledgment of the violence inherent in this portrayal, demands a larger readership. Viewing the history of the United States through the vein of its imperialist, nineteenth-century expansionist goals, as well as through the lens of a Mexican American woman's historicist maneuvers, is to incorporate a more nuanced understanding of hemispheric American spaces into the classroom. Studying the life and work of González provides a much-needed starting point for discussions of Mexican and American relations as well as the geographic perspective of the Southwest within American literature and history classrooms.

Notes

1. We wish to thank Caroline Levander for introducing us to González, as well as for her kind inclusion of our essay in this collection and for her generous assistance in its development at every stage. Thanks also go to Alexander X. Byrd, Anthony B. Pinn, and Michael O. Emerson for their helpful comments during the development of this essay. We are also grateful to the Humanities Research Center at Rice University for providing funding to visit the E. E. Mireles and Jovita González de Mireles Papers, Special Collections & Archives, Texas A&M University-Corpus Christi, Mary and Jeff Bell Library, TX; Edmundo E. Jovita González Mireles Papers, Southwestern Writers Collection, Alkek Library, Texas State University, San Marcos, TX; and Texas Folklore Society Records, 1909–1970, Center for American History, The University of Texas at Austin.
2. González and Eve Raleigh, *Caballero: A Historical Novel* (College Station, TX: Texas A&M University Press, 1996). In addition to *Caballero*, several modern editions of Gonzalez's work that are discussed in this essay have been recently published: *Dew on the Thorn* (ed. José Limón [Houston: Arte Público Press, 1997]); *Life Along the Border: A Landmark Tejana Thesis*, Gonzalez's master's thesis (ed. María Euginia Cotera, Elma Dill Russell Spencer Series in the West and Southwest [College Station, TX: Texas A&M University Press, 2006]); and *The Woman Who Lost Her Soul and Other Stories*, a collection of Gonzalez's folklore tales (ed. Sergio Reyna, Recovering the U.S. Hispanic Literary Heritage series [Houston: Arte Público Press, 2000]). All future citations will refer to these editions.
3. Cotera, "A Woman of the Borderlands: 'Social Life in Cameron, Starr, and Zapata Counties' and the Origins of Borderlands Discourse," Introduction to *Life Along the Border: A Landmark Tejana Thesis*; Cotera, "Jovita González Mireles: A Sense of History and Homeland" in *Latina Legacies: Identity, Biography, and Community*, eds. Vicki Ruiz and Virginia Korrol (New York: Oxford Univ. Press, 2005); Cotera, *Native Speakers: Ella Deloria, Zora Neale Hurston, Jovita González, and the Poetics of Culture* (Austin: Univ. of Texas Press, 2008); Limón, *Dancing with the Devil: Society and Cultural Poetics in Mexican-American South Texas* (Madison, WI: Univ. of Wisconsin Press, 1994).
4. Leticia Magda Garza-Falcón, *Gente Decente: A Borderlands Response to the Rhetoric of Dominance* (Austin: University of Texas Press, 1998), 8. Louis Gerard Mendoza also argues for a Chicano reading of *Caballero*, stating,

"[w]hat *Caballero* provides us with is a mechanism for imagining the conditions and form of an emerging oppositional consciousness in this transitional period" (*Historia: The Literary Making of Chicana and Chicano History* [College Station: Texas A&M University Press, 2001], p. 55.
5. See, for example, David Gutierrez, "Significant to Whom? Mexican Americans and the History of the American West," *Western Historical Quarterly* 24.4 (Nov. 1993): 519–539; B. J. Manríquez, "Argument in Narrative: Tropology in Jovita González's *Caballero*," *Bilingual Review/La revista bilingue*, 25:2 (2000): 172–178; J. Javier Rodríguez, "Caballero's Global Continuum: Time and Place in South Texas," *MELUS: The Journal of the Society for the Study of the Multi-Ethnic Literature of the United States*, 33.1 (2008 Spring): 117–138; and Andrea Tinnemeye, "Enlightenment Ideology and the Crisis of Whiteness in *Francis Berrian* and *Caballero*," *Western American Literature*, 35:1 (2000): 21–32.
6. John Major, *Prize Possession: The United States Government and the Panama Canal 1903–1979* (Cambridge: Cambridge Univ. Press, 2003) p. 68; M. G. Bulmer, *Francis Galton: Pioneer of Heredity and Biometry* (Baltimore: John Hopkins Univ. Press, 2003) p. 84; and Mark Dyreson, "The 'Physical Value' of Races and Nations: Anthropology and Athletics at the Louisiana Purchase Exposition" in *The 1904 Anthropology Days and Olympic Games: Sport, Race, and American Imperialism*, ed. Susan Brownell, Critical Studies in the History of Anthropology Series (Lincoln, NE: Univ. of Nebraska Press, 2008), p. 142.
7. Dick D. Heller, Jr., "Roma-Los Saenz, Texas" (2008) in *The Handbook of Texas Online*, http://www.tshaonline.org/handbook/online/articles/RR/hgr6_print.html (accessed 6 May 2009). For additional information on Roma, see Benjamin Heber Johnson, *Bordertown: The Odyssey of an American Place* (New Haven: Yale University Press, 2008).
8. Cotera, "A Woman of the Borderlands," 3–33, 6–9.
9. David Montejano, *Anglos and Mexicans in the Making of Texas, 1836–1986* (Austin: Univ. of Texas Press, 1987), p. 160. For a study on the history of Mexican American educational struggles see Guadalupe San Miguel Jr., *"Let All of Them Take Heed": Mexican Americans and the Campaign for Educational Equality in Texas, 1910–1981* (Austin: Univ. of Texas Press, 1988), pp. 19, 23–24, 51, 54. See also Arnoldo De León, *They Called Them Greasers: Anglo Attitudes Toward Mexicans in Texas, 1821–1900* (Austin: University of Texas Press, 1983).
10. Debbie Mauldin Cottrell, "Casis, Lilia Mary (1869–1947)" (2008) in *The Handbook of Texas Online*, http://www.tshaonline.org/handbook/online/articles/CC/fcace_print.html (accessed 6 May 2009).
11. Cotera, *Latina Legacies*, pp. 158–174, 161–164.
12. Limón, *Dancing with the Devil*, p. 45.
13. While it is unclear whether Dobie studied under Boaz at Columbia, the two corresponded and met throughout the 1910s and 1920s, during which Dobie was an active member of the American Folk-Lore Society and Boaz

was publications editor for the *Journal of American Folk-Lore*. Their tense correspondence came to an end in 1925; see Paul Stone, "J. Frank Dobie and the American Folklore Society" in *Corners of Texas*, ed. Francis Edward Abernathy (Dallas: University of North Texas Press, 1993), pp. 47–66.
14. Nolan Porterfield, *Last Cavalier: The Life and Times of John A. Lomax, 1867–1948* (Chicago: Univ. of Illinois Press, 2001), p. 141.
15. Francis E. Abernethy, "Dobie, James Frank (1888–1964)" (2008) *The Handbook of Texas Online*, http://www.tshaonline.org/handbook/online/articles/DD/fdo2.html (accessed 21 May 2009).
16. González, Untitled Handwritten Autobiography, Texas A&M University-Corpus Christi.
17. Folklore Society of Texas Bulletin, 1911, The University of Texas at Austin. For more information on the early history of the Texas Folklore Society, see Francis Edward Abernethy, *The Texas Folklore Society*, 51 vols. (Denton: University of North Texas Press, 1992), vol. 1.
18. Folklore Society of Texas Bulletin.
19. Texas Folklore Society Records, 1909–1970. After examining the society's business and general records, we discovered very little information tying Gonzalez to the operations of the organization beyond her presentations at meetings and occasional listing as an officer on the programs of the society's annual gatherings. It is also clear from these records that Dobie shouldered much of the responsibility for the organization, as evidenced by his consistent listing as an officer, most often secretary-treasurer, on the society's annual programs.
20. Gonzalez's evolving belief in the submerged politics of ethnographic writings of the TFS are tantalizing, suggested by the titles of her own publications, as well as the politics of their plots. For example, in addition to the more typical work on "The Folklore of the Texas-Mexican Vanquero" (*Publications of the Texas Folklore Society* 4 [1927]: 7–22) and "Tales and Songs of the Texas-Mexicans" (*Publications of the Texas Folklore Society* 8 [1930]: 109–114), Gonzalez also published "America invades the Border Towns" (*Southwest Historical Review* 15 [1930]: 469–77) and "Among My People" (*Publications of the Texas Folklore Society* 8 [1930]: 99–108).
21. González, Untitled Handwritten Autobiography, Texas A&M University-Corpus Christi.
22. Abernethy, "Dobie, James Frank (1888–1964)" (2008), in *The Handbook of Texas Online*.
23. William C. Pool, "Barker, Eugene Campbell (1874–1956)" (2008), *The Handbook of Texas Online*, http://www.tshaonline.org/handbook/online/articles/BB/fba65.html (accessed 21 May 2009). For further information on Barker, see William C. Pool, *Eugene C. Barker, Historian* (Austin: Texas State Historical Association, 1971).
24. González, Untitled Handwritten Autobiography, Texas A&M University-Corpus Christi.

25. González, Untitled Handwritten Autobiography, Texas A&M University-Corpus Christi.
26. González, Untitled Handwritten Autobiography, Texas A&M University-Corpus Christi.
27. Cotera, "A Woman of the Borderlands," 14.
28. David H. Stevens to Miss Josita Gonzles (*sic*), 5 June 1934, Texas A&M University-Corpus Christi.
29. Raymond Fosdick, *The Story of the Rockefeller Foundation* (New York: Harper & Row, 1952), p. 239.
30. Fosdick, p. 242.
31. González, Untitled Handwritten Autobiography, Texas A&M University-Corpus Christi.
32. Cotera, Latina Legacies, 172.
33. J. Frank Dobie to Mrs. Jovita González de Mireles, 27 April 1937, The University of Texas at Austin.
34. Texas Folklore Society Records, 1909–1970; Cotera, *Latina Legacies*, pp. 168–173. The extent to which *Caballero: A Historical Novel* is a collaboratively written text is still a matter of much debate. Cotera's reading of *Caballero* is strengthened by the frequent silences in the archival holdings, which often provides a more nuanced presentation of Mireles's and Eimer's thoughts than González's own in order to emphasize the collaborative aspects of the novel. Cotera's meticulous attention to the textual and biographical relationship between González and Eimer leads her to read *Caballero* as a wholly collaboratively produced novel whose joint origin bleeds into the plot, in which "women, queers, and peons... develop new strategies for survival both for themselves and for their nation" (*Latina Legacies*, p. 451).
35. Jovita González de Mireles to Dr. John Joseph Gorrell, 26 August 1939, Texas A&M University-Corpus Christi.
36. Jovita González de Mireles and Margaret Eimer (Eve Raleigh), 30 May 1939, Texas State University, San Marcos.
37. D. Thorne to Miss de Mireles, 21 August 1939, Texas A&M University-Corpusssti.
38. M. E. to Jovita, n.d., Texas A&M University-Corpus Christi.
39. This point is also made by Pablo Ramirez, "A Borderlands Response to American Eugenics in Jovita González and Eve Raleigh's *Caballero*," *Canadian Review of American Studies*, 39.1 (2009): 21–39, 27.
40. The tensions stemming from González's own struggles with race and class are most evident in the narrator's stance toward the indigenous Native Americans in Texas. Oddly, the narrator's approach to Native Americans is static, regardless of the change in national politics, and is silent about the racism directed toward them by nearly every character: A priest notes that Mexico gave the rancheros land "beset as it was with marauding Indians," and Don Santiago argues that his family owns the land by right because they have "held against the Indians for a hundred years" (*Caballero*, p. 56,

50). Don Santiago also notes that Alvaro's energy, after failing to be spent on chasing Rangers along the border, can be spent on "the Indians and cattle thieves" (*Caballero*, p. 141). The troubled racial politics of the novel are perhaps most glaring in an instance of what is not mentioned, rather than what is: Within the text, the narrator inexplicably makes no attempt at representing African Americans or discussing the looming U.S. Civil War.
41. De León, *They Called Them Greasers*, p.76; Montejano, *Anglos and Mexicans*, pp. 33–34.
42. Exploring the architectural and ethnographic spaces of the hacienda presented throughout the novel, Monika Kaup persuasively argues that "*Caballero* unmasks the class-based ideology of hidalgo landownership underlying Mexican frontier nationalism, arguing that nationalism cannot claim to be the universal voice of Mexican resistance against U.S. colonization because it seeks to preserve feudal structures of domination at home" ("The Unsustainable *Hacienda*: The Rhetoric of Progress in Jovita González and Eve Raleigh's *Caballero*," *Modern Fiction Studies*, 51:3 [2005]: 561–591, 571).
43. Jovita González de Mireles to Dr. John Joséph Gorrell, 26 August 1939, Texas A&M University-Corpus Christi, Corpus Christi.
44. Jovita González de Mireles to Dr. John Joséph Gorrell, Texas State University. Vincent Pérez explores the cinematic and literary connections between *Caballero* and *Gone with the Wind*, as well as *Caballero*'s larger redeployment of Southern cultural history, in "Remembering the Hacienda: History and Memory in Jovita González and Eve Raleigh's *Caballero: A Historical Novel* (in *Look Away! The U.S. South in New World Studies*, eds. John Smith and Deborah Cohn, New Americanists series [Durham, NC: Duke University Press, 2004], pp. 471–494).
45. Sylvia Ann Grider, "Dorothy Scarborough," in *Texas Women Writers: A Tradition of Their Own*, eds. Sylvia Ann Grider and Lou Halsell Rodenberger (College Station, TX: Texas A&M University Press, 1997), pp. 134, 137.
46. See K. Jack Baur, "Mexican War" (2008), *The Handbook of Texas Online*, http://www.tshaonline.org/handbook/online/articles/MM/qdm2.html (accessed 6 May 2009) for further information.
47. Cotera, "Woman of the Borderlands," pp. 17–18.
48. See Sommer, *Foundational Fictions: The National Romances of Latin America* (Berkeley: Univ. of California Press, 1993) pp. 1–51.
49. Pérez, p. 474.
50. Montejano, pp. 34–35.
51. Alonzo, *Tejano Legacy: Rancheros and Settlers in South Texas, 1734–1900* (Albuquerque: University of New Mexico Press, 1998), p. 114.
52. John González, "Terms of Engagement: Nation or Patriarchy in Jovita González's and Eve Raleigh's *Caballero*," *Recovering the U.S. Hispanic Literary Heritage*, eds. José F. Aranda Jr. and Silvio Torres-Saillant, 5 vols. (Houston: Arte Público Press, 1993), 4: 264–276, 268.

CHAPTER 10

Migrant Archives: New Routes in and out of American Studies

Rodrigo Lazo

The history of the modern archive is inextricable from the establishment of nation-states. In various parts of the world, including France in 1790 and numerous countries in the nineteenth and twentieth centuries, the establishment of a "national archive" followed a revolutionary break from monarchy or colonialism. Archive and nation came together to grant each other authority and credibility: One contained documents and records that supposedly spoke to and about the state, while the nation granted a certain cachet to an archive, elevating it above its local and regional counterparts. The continuing influence of that institutional formation is evident in a statement produced by the International Council on Archives, a professional society for archivists and their institutions: "Archives constitute the memory of the nations and of societies, shape their identity, and are a cornerstone of the information society."[1] The high stakes of such a claim begin to explain the sustained examination and critique of archives by theorists, most prominently, Jacques Derrida, as well as historians, librarians, and other scholars.[2] If archives do indeed "constitute" memory rather than just contain it or record it, and if they are crucial in disseminating information, a variety of questions emerge. How do archives develop procedures for the inclusion and exclusion of materials, for the preservation and even inadvertent destruction of information? How do archives wield authority over what is considered important in public institutions and educational settings? Who has access to archives, and what types of identity claims are made by the people who control and disperse the information? Most pertinent to the goals of this essay, how can we conceive of archives in new ways so that research agendas are no longer contained by the parameters of the archival frame?

The physical manifestation of archival parameters in the modern period has been a building, which literally housed the materials and served metaphorically to delimit the information. The records office or rare-book library provided safekeeping for documents and became a place where archive rats could gather empirical evidence for their accounts of the past.[3] In the United States, the National Archives occupy both a modern research facility in College Park, Maryland, and a museum facing Constitution Avenue in Washington, D.C., not far from the Capitol. The latter location features copies of the Declaration of Independence, the U.S. Constitution, and the Bill of Rights.[4] This tourist attraction, resembling the Parthenon and featuring New Deal–era engravings on its façade, provides an image of the nation that is connected to a grand narrative of freedom stemming from 1776. The main chamber, where the Declaration of Independence and other documents are displayed, is known as the Rotunda for the Charters of Freedom. By contrast, the facility in College Park, opened in 1994, is a research center that holds many of the records of government agencies. Although some records are still in the Washington, D.C., location, the separation into buildings known as Archives I and Archives II in effect divides visitors into tourist and researcher.

But that division is breaking down as the archive building shares space with the archive webpage. With rare documents, including everything from nineteenth-century immigration records to seventeenth-century books, increasingly becoming available through online databases, archives are now accessible to someone in his or her house, a potential alternative (and more classical) site for archive construction. Visitors to the U.S. National Archives webpage can now "Collect records to create your own archive of American history."[5] A personal archive has the potential to challenge the authority of the national building. And scholars who conceptualize archives in ways that displace the terms under which a singular archive is constructed can open new routes for research agendas.

For American studies, these new routes allow for a way to move in and out of the nation rather than privileging national study as the defining point of the field. It is my contention that these paths will lead us to what I call "migrant archives." Migrant archives reside in obscurity and are always at the edge of annihilation. They are the texts of the past that have not been written into the official spaces of archivization, even though they weave in and out of the buildings that house documents.

Migrant archives are oxymoronic because one of the functions of archival organization is storage in a specific (safe) place. Migration, by contrast, connotes a journey. And "migrant archives" call for a journey, if not on the part of the researcher, then certainly on the part of a text that will have to travel from one place to another. The contemporary association of migration with border crossings and movement of people in and out of nation-states at great personal risk emphasizes that migration is not safe. Certainly, it is at odds with a storage vault, which historically has been one of the constituting elements of an archive.

The move toward migrant archives calls for the ongoing examination of how memory is constituted, how history is written, and how research is connected to identity. In other words, control of the archive has epistemological and political ramifications. For American studies to move beyond the fixed archive of an Anglo-American nation, scholars will have to undertake more multilingual work in migrant archives. Texts written in languages other than English can lead scholars to alternative ways of remembering the past, new ways of naming multiple nations and communities, and even the invention of new ontologies. The writing itself—in Spanish, for example—can constitute a different archive that calls attention to the political choices of writers and also implies that they were unable or unwilling to write in English. More than a record of social processes or a representation of experiences, the writing itself, whether in book form or scraps of paper, is a site where migrant conditions take material form. As writing moves to the de facto official language of a discipline or area, translation becomes a particularly important but unpredictable and vexing type of work. Translation can integrate marginalized and forgotten people into the authorized archive even as it threatens to alter the content of a migrant archive and erase, however gently, language differences. Given the importance of translation, I begin my discussion by focusing on *Mis Memorias,* a book whose recent republication involved movement out of a migrant archive across decades and languages. I pair this with a return to Derrida's *Archive Fever,* a fruitful critique of the archive that has inspired lively responses. I conclude with a consideration of how migrant archives emerged from the project of the Recovering the U.S. Hispanic Literary Heritage.

In 1935, Luis G. Gómez published *Mis Memorias*, a Spanish-language memoir that recounted Gómez's experiences crossing the border from Mexico to the United States and working as an accountant and contractor in Southeast Texas. For many years, the few known remaining copies of *Mis Memorias* were kept by Gómez's descendants but were not circulated among the general public. In 1991, Gómez's grandson presented information about the book at a meeting of the Spanish American Genealogical Association in Corpus Christi, Texas. That presentation drew the attention of Thomas H. Kreneck, associate director for special collections and archives at Texas A&M University, Corpus Christi; Kreneck found that the book was not listed in library catalogues. Working with Gómez's family, Kreneck helped usher the book into publication in its translated form, *Crossing the Rio Grande*, a book out of a migrant archive that crossed decades, generations, and languages.[6]

Mis Memorias resided in a migrant archive, not only because it was kept privately and faced the risk of being lost or destroyed. It also had another mark of migrancy in the U.S. context: It was written in a language other than English. The editors of the book attempt to counter the lack of attention to the Spanish language by undertaking two archival functions: the establishment of a lineage for the book (in this case, one based on a family) and the storage of information in an official repository. In his introduction to the volume, Kreneck describes the process of translation and republication as a "labor of love and family devotion." He characterizes the book as "truly a product of two men," Gómez and his translator, Guadalupe Valdez, who was also Gómez's grandson. "By making these memoirs available to a wider audience, Valdez has done much not only for his grandfather's memory and for his family's heritage but for scholarship as well," Kreneck writes.[7] Like Kreneck, Valdez emphasizes the importance of family lineage while explaining why and how he took on the project: "Because of my special relationship with my grandfather and because I am the last living member of his family who knew him personally, it has naturally fallen to me to translate his memoirs."[8] In claiming a "natural" reason for undertaking such work, Valdez conflates academic labor and genealogical connections. That perspective celebrates the regenerative potential of family history even as it overlooks the responsibility of professional scholars to do such work. Why shouldn't professors be engaged in that type of recuperative republication and translation? In placing familial ties at the center of the archival process, Kreneck and Valdez actually turn to a classical notion of the archive, one that linked the archive to the archivist's house.

The family, connected to living quarters and offering a genealogical connection from the present to the past, is an important dimension of the archive. The etymology of the word "archive," as Jacques Derrida noted, goes back to an association in ancient Greece between records and a house run by a patriarch known as an *archon*. Derrida writes,

"As is the case for the Latin *archivum* or *archium* (a word that is used in the singular, as was the French *archive*, formerly employed as a masculine singular: *un archive*) the meaning of 'archive,' its only meaning, comes to it from the Greek *arkheion*: initially a house, a domicile, an address, the residence of the superior magistrates, the *archons*, those who commanded."[9] A*rkhe* can be translated as a commandment. By this account, the father's position in the house (and his commandment) was linked to the maintenance of records. In the modern period, the rise of the nation-state transfers the *archon*'s authority to public institutions, and not just national archives, but also regional historical societies and local records offices. The public repository continues to imply a common legacy, with the nation or local community taking the place of the family. And thus it is no surprise that the editors of *Crossing the Rio Grande* are overt in communicating the relationship developed by Gómez's descendants with various archival establishments.

As *Crossing the Rio Grande*'s introductory material makes clear, one of the goals of the publication is to situate the book in various types of archives. *Mis Memorias/Crossing* moves into a rare-book repository, a card catalogue, and even a field of study. In conjunction with the new publication, the family bestows a copy of the Spanish-language edition to Texas A&M library, to the delight of Kreneck, the archivist. The English-language edition becomes part of another archive, the library catalogue; libraries across the country take on the archival function of organizing and legitimating published material. The book enters a third archival space through the participation of a university press. Published by Texas A&M in an attractive hardcover edition, *Crossing the Rio Grande* has the potential to move into the archive created by disciplines and fields of study. The book can become part of the reading material of American studies, Texas history, labor studies, and Hispanic literary heritage, among other fields.

My usage of "archive" here is in keeping with a pronounced slippage in the meaning of the word in recent years. Once invoked in certain disciplines to distinguish repositories of rare documents from libraries, "archive" is now used in reference to a record of web postings, historical memory, libraries, and even a set of readings with a theme, among others. Scholars in American studies regularly now refer to an "alternative archive" that will help the field reconsider its assumptions and practices for selecting texts. Two related forces are at work here, writes Marlene Manoff, "One is the conflation of libraries, museums, and archives; and the other is the inflation of the term 'archive,' which has become a kind of loose signifier for a disparate set of concepts."[10] The term is being used more widely, I would argue, because the physical archive is associated with authority and is a locus of power in research. Calling a selected group of readings "my archive" claims legitimacy but can also invoke the types of critiques that have been leveled at archive, the building. Perhaps writers use the term ironically. But does a site that recovers and collects texts for public use justifiably stake a more serious claim on the use of "archive" than individual researchers who use the word loosely? When I say that *Crossing the Rio Grande* enters several archives, it is to note that the library and university press can grant visibility in academic or other public discourse to what might otherwise remain a family heirloom.

Crossing the Rio Grande enters various archives because the new English edition has the potential to reach a reading audience that includes monolingual scholars. Translation becomes an important process in the recovery of migrant archives. Like migrancy, translation implies movement; literally, "to carry across." In the case of American studies, the movement in and out of migrant archives calls for the transferring of little-known documents into the more visible spaces of the field's debates and also for the carrying of those documents across languages. In the United States, archives hold a wealth of textual riches in languages other than English. A decade ago, Marc Shell and Werner Sollors attempted to "make visible the most glaring blind spot in American letters" by publishing *The Multilingual Anthology of American Literature*, which included selections in French, Navajo, and German, among other languages.[11] This anthology invoked literary history as a kind of memory that could intervene in contemporary debates about the United States' relationship to English-only movements. Multilingualism calls for an opposition to the national fixation on a single language, and it opens avenues for transnational connections. Materials in languages other than English written in the past face the possibility of disappearance and annihilation because historically, the United States has not established official channels for study and archivization. With the din of a call for English as an "official language" persistently in the background, the papers of multilingual America remain in the obscurity of archival vaults, assuming that they have been gathered and kept. In addition, the demands of working in multiple languages means that relatively few scholars can approach the multilingual past; a project for a multilingual America would involve extensive collaboration. It is difficult enough to inspire bilingual approaches. Because many, if not most, practitioners of American studies work predominantly or exclusively in English, documents require translation, which is always intertwined with interpretation, in order to enter the field. In other words, translation helps formulate a recognizable body of work that the field can share, a point I take up in more detail later. As the case of *Crossing the Rio Grande* illustrates, translation also becomes part of the process of building archives (field of study and library holding).

But how does a field of study resemble a repository of documents? What exactly is the relationship between American studies and the U.S. National Archives? That question is an important dimension of Derrida's *Mal d'archive*, which seeks to posit a homology between the field of psychoanalysis and the archive proper. With a founding father (Freud) who bestows authority on the operations of interpretation, psychoanalysis offers an institutional narrative that is comparable to that of the national or local archive. Derrida argues that in both cases the archive posits an origin (a primal scene, the founding of the nation) and frames a narrative based on that claim. As such, Derrida seeks to undo the ontology of the archive.

Mal d'archive runs through a gamut of associations, connecting the archival setting with everything from politics to e-mail. Derrida's trail links the following: the aforementioned ancient Greek house as repository of archival material, the keeper of the archive as a kind of father or authority, the institution as a protector of the archival function, an academic and scientific field of study as an archival impression, and the storage of material as a process of violent exclusion. In turning to psychoanalysis, Derrida attempts to combat the privileging of origins and familial connections. At the same time, his broad strokes turn the archive into a question of memory, both personal and collective. The Derridean dispersal of the archive differs significantly from Michel Foucault's use of "archive." In *The Archeology of Knowledge*, Foucault presents the "archive" as a discursive system that permits certain things to be said.[12] By contrast, Derrida would grant no such presence to the "archive," and his critique helps us unpack the relationship (and distinction) between archive as a place with research materials and reading desks, and archive as metonym for the organization of information.

To develop his critique of official sources of knowledge, Derrida turns to Yosef Hayim Yerushalmi's *Freud's Moses: Judaism Terminable and Interminable* and thus intertwines history (Yerushalmi's discipline) with Judaism and psychoanalysis. Much of Derrida's discussion leads to a confrontation with Yerushalmi's question: Is psychoanalysis a Jewish science? Descent and legacy, to which I have already alluded in relation to *Mis Memorias/Crossing the Rio Grande*, would seem constitutive elements in both psychoanalysis and Judaism. One relies on biology and the other on the name of Freud as an intellectual father, but Derrida also calls attention to how the ethnos and father come together in a field of study, and that has ramifications for the construction of an ethnic archive. Challenging this association, Derrida writes,

> But the structure of the theoretical, philosophical, scientific statement, and even when it concerns history, does not have, should not in principle have, an intrinsic and essential need for the archive, and for what binds the archive in all its forms to some proper name or to some body proper, to some (familial or national) filiation, to covenants, to secrets. It has no such need, in any case, in its relationship or in its claim to truth—in the classical sense of the term.[13]

The point here is that language, a statement linked to a proposition, does not need an archive to stake its claim on truth. In other words, a scientific or philosophical finding would not need Judaism or Freud for justification. For our purposes, we could say that the textual record, the scraps that might emerge from migrant archives, does not need "some body proper" (an author) or some place (a national archive) for the validation that we would associate with a claim to truth. More concretely, Luis G. Gómez should not need the archival setup established by his descendants. But is that actually the condition of scholarly work in the United States at this time? Does a book such as *Crossing the Rio Grande* actually need filiation in order for its claims to enter a public discussion? Does it need a home for that claim to be heard and, if so, what type of archive?

Derrida would allow no such filiations. The stakes of his argument against the archive become clear when, deploying the language of psychoanalysis, he associates the archival function with the death drive. On the one hand, the drive to establish archives, archive fever, is related to a kind of conservation and preservation. But according to Derrida, it also erases what came before it. The effects of the death drive are not confined to the exclusionary effect of collecting some materials and not others in an archive. Rather, Derrida posits that the archival process releases a type of aggression; the archive is an impression that alters a previous impression. More than excluding something, the archive destroys the archive that preceded it. One need not go far here to come up with examples for Derrida's theoretical point. Surely, the preservation of English as a kind of official language has led to the annihilation of American Indian languages. With English as a language of America, the country or hemisphere's more ancient copy is subjected to erasure.

The association of archival construction with the death drive begins to explain Derrida's provocative line that "archive fever verges on radical evil."[14] From a Kantian perspective, the notion of "radical evil" refers not to an incomprehensible or momentous example of evil but rather a kind of conduct that proceeds only from personal inclination or interest. Radical evil is not about committing atrocities, no matter how heinous. It is about acting in a way that discounts the possibility of a higher principle.[15] The Derridean critique, then, would appear to rest on the sense of the archivist as someone who conforms to the structure of the archive for some benefit, which could be personal but not necessarily imply malicious conduct. One might consider here the pleasure of hoarding books or the patriotic feeling of running a national repository as self-serving personal rewards that also create the boundaries of authoritative thinking. In effect, archive collections can inhibit avenues of research that might move closer to alternate truth claims.

The introduction of an ethical dimension calls attention to the different associations created when *Mal d'archive* (1995) was translated into the English *Archive Fever* (1996). The French *mal* connotes not only a kind of illness but also can be synonymous with "evil." As a connotative sign, it is more varied and rich than "fever." If anything, the translation of the title of Derrida's book into English opts for a satirical effect. "Archive fever" sounds like a kind of fad, perhaps a headline in a newsweekly; it is reminiscent of the 1970s song "Boogie Fever." The translated title emphasizes a critique of the archival turn as a form of empirical research and thus differs from the engagement of *mal* with European philosophy. With the focus on Yerushalmi, a historian, and the use of "fever," it is not surprising that Derrida inspired responses from researchers who are invested in archival work: historians.

In her critique of Derrida, Carolyn Steedman calls attention to the material realities of labor (within and without the archive), at times connecting the physical experience of going to and working in the archive with the bodily effects of production. Running with Derrida's fever metaphor, Steedman notes that sometimes the dust in archives can make people sick. More importantly, dust is itself a metaphor for the residue of the laboring classes. Reading the historian Michelet alongside Derrida, Steedman calls attention to "the dust of the workers who made the papers and parchments; the dust of the animals who provided the skins for their leather bindings."[16] Michelet, she notes, "inhaled the by-product of all the filthy trades that have, by circuitous routes, deposited their end-products in the archives."[17] As one who actually goes to archives, Steedman shifts the focus away from the Derridean concern with the archive as a locus of authority in language. In other words, the archive can also contain information about those who do not have authority, and even those whose authority has been forgotten. And yet Steedman's critique does not negate Derrida's point that the archive does claim an authority, usually on a foundation stemming from a nation, a locality, an institution, or a person. This locus of authority becomes the familiar point that frames the archival holding, including the migrant archives that might be contained within.

Making a book familiar is one of the goals of *Crossing the Rio Grande*. In moving the family book into the public space of libraries and doing so in standard English, Gómez's descendants desire to move an ancestral authority into an official archive. The translation of the book's title is important and, in some ways, more radical than the change from "*mal*" to "fever." "*Mis Memorias*" emphasizes the subjective and very personal claim made by its author. A literal translation, "my memories," does not have the right intonation, but other possibilities, "a memoir" or "my experiences," would have retained the generic and personal emphasis. The title *Crossing the Rio Grande*, by contrast, chooses a common trope that describes the passage from Mexico to the United States and thus situates the book in a public discussion of immigration and the experiences of those who make the trek across that angry river. Here we see the importance of translation to the archival function because the title helps situate the book in the archive of immigration. But the change also prompts a question about how translation might be the type of act that Derrida calls "radical evil." I emphasize that evil here is not about heinous acts. But if Derrida's use of "radical evil" speaks to an automatic following of the norm, that which is accepted, then displacing the Spanish original with a translation into the official language does negate the presence of the Spanish language in U.S. publication history.

Perhaps that new impression, with its potential to obliterate, is an inevitable effect of translation, a point that has inspired lengthy discussion in translation studies. Lawrence Venuti, for example, has written, "Translators are very much aware that any sense of authorial presence in a translation is an illusion, an effect of transparent discourse, comparable to a 'stunt,' but they nonetheless assert that they participate in a 'psychological' relationship with the author in which they repress their own personality."[18] Gómez's translator takes that relationship to a grand-Oedipal degree, simultaneously asserting a blood connection to the author, a claim to the author's presence (without an acknowledgement of illusion), and a foregrounding of his own presence as descendant-translator. The translator-grandson thus carries across the book from migrant archives to a singular archive that demands a standard language. That integration of language difference into a norm (English) is one reason why scholars of hemispheric American studies have increasingly called for attention to translation studies and sought new ways to conceive of the transformation created by translation, including "adaptation."[19]

Despite the effects of the impression created by the new English version of *Mis Memorias*, something is also gained by bringing this document out of migrant archives. Given the anti-immigrant sentiments that surface repeatedly in U.S. history, the translation and public dissemination of the Gómez memoir is a necessary evil. In other words, the evil is idiomatic more than ethical. The necessity here is about political participation in U.S. society. In a footnote early in his book, Derrida gestures to an important claim that he does not follow. He writes, "There is no political power without control of the archive, if not memory."[20] Here Derrida reminds us that archives are also part of a political arena in which people vie for power and recognition. If we take this statement as a reference to the archive as a repository of historical documents (although the usage there is more diffuse), questions emerge about the relationship between research emphasis and the exercising of power in a contemporary liberal society. Derrida's point would then be that political power derives in part from such a national institution because the archivist interprets meanings and associations. In some ways, the archive defines the nation, and participation in the archive is one gauge of democratization. Still, Derrida's sentence raises an additional problem, that of memory. "Nul pouvoir politique sans contrôle de l'archive, sinon de la mémoire."[21] The phrase "if not memory," not quite conditional and almost an afterthought, adds a complication. Is it national memory? Personal memory? Your memory? Given the subjective thrust of memory, the point would be that political power rests on the control of the memories of others. When *Mis Memorias* is translated into *Crossing the Rio Grande*, Gómez's account becomes part of a different archive, one that is in dialogue with the recent resurgence of anti-immigrant discourse. It is not surprising that *Crossing the Rio Grande*'s editors emphasize the importance of immigrant labor, and thus the book's subtitle, "An Immigrant's Life in the 1880s." This type of book is a necessity in a society that routinely seeks to forget its immigrant past and its multilingual background.

And yet the archive of immigration, with its monument at Ellis Island and reading lists of immigrant novels, would seem to bind some of the energy of *Mis Memorias*. An "immigrant life" has traditionally implied integration or even assimilation into a national panorama. The translation integrates Gómez's book into an archive rather than emphasizes the multiple locations of the book's publication history and its content. *Mis Memorias*'s episodes recount movement within Texas and note that returning to Mexico is an important consideration for the author and his friends. The change into an immigrant life de-emphasizes what may be the most interesting dimension of the book's social panorama: the importance of a Spanish-speaking workforce that is not easily assimilated into Texas society in the late nineteenth century.

The challenge not just for American studies but also for many scholarly efforts that recover texts from the past is how to bring forth historical and linguistic differences without allowing the present discourse to dominate. The archival claim, meaning the terms under which an archive is constructed, always threatens to become hegemonic, but some texts may contain a difference emphatic enough to prompt a reconsideration of the archive's limits. Let me be more specific by looking at the case of a very important archive developed in the last fifteen years: the Recovering the U.S. Hispanic Literary Heritage Project.

The "Recovery Project" is an example of an archive in that it collects, houses, and circulates a variety of documents under the rubric of ethnic identity. The project is authoritative, in the sense that Derrida uses the term, because it gains legitimacy from describing its object of study as the textual record left behind by people of Hispanic descent. Although the project is capacious in its definition of "Hispanic" and is also sensitive to the variety of populations that might be part of that group historically, the claim to heritage retroactively organizes the project's texts under the rubric of a contemporary identity formation. One of the salutary effects of such an archive construction is that the Recovery Project has brought into circulation materials that were previously available only in the rarest of library stacks. It was the Recovery Project that in the 1990s published María Amparo Ruiz de Burton's novels *Who Would Have Thought It?* (1872) and *The Squatter and the Don* (1885) as well as the anonymous *Jicoténcal* (1826). Housed at the University of Houston, the Recovery Project now holds originals, photocopies, and microfilms of thousands of documents, thus functioning both as a repository of rare materials and as an institutional home.[22]

The goal in establishing the Recovery Project was not only to identify the Hispanic past but also to restore it, presumably as an important component of the nation, if not memory. This was, like so much of American literary revision in the 1980s and 1990s, an attempt to broaden the texts that made up a literary past. In the introduction to the first volume of the Recovery series, Ramón A. Gutiérrez and Genaro M. Padilla write, "The long forgotten editorials these men and women wrote, their manifestoes for better wages and better working conditions, their private thoughts and emotions committed to diaries, their moral tales disguised as comedies and farces, their tersely measured lyrical poems, and their pauses and silence in the textual record are the collective object of our study." This expansion beyond literary genres and the public sphere of newspapers and pamphlets moved interdisciplinary scholarship into "private thoughts and emotions," the terrain of memory and lived experiences. In some cases, the "literary" would give way to Hispanic heritage, defined as the cultural background of "these men and women." In working against the "long forgotten," the memory lapses and the destruction, the Recovery Project could simultaneously make visible a documentary history of a people.[23] a proper endnote. As far as I can tell the information is: Gutiérrez and Padilla, "Introduction," Recovering the US Hispanic Literary Heritage (Houston: Arte Público Press, 1993), 21.]

Despite what we might call the archival threat of this scholarly effort, the project was a great necessity in that it promoted a type of research that was not being carried out on a large scale. More than a century before the Recovery Project, Walt Whitman, as Levander has noted, articulated, on commemorating the 333rd anniversary of the settlement of Santa Fe, New Mexico, a need to consider the Hispanic past: "We Americans have yet to really learn our own antecedents, and sort them, to unify them."[24] Calling on the nation to move beyond its Anglocentrism, Whitman argued, "To that composite American identity of the future, Spanish character will supply some of the most needed parts."[25] Whitman's statement calls attention to the ongoing need for recovery projects of various sorts, and it also poses one of the problems of an authoritative archive. In the Whitmanian schema, the search for a Hispanic antecedent would recover parts of a "composite" U.S. identity; the goal would be unification into what we might call a multicultural nation. By contrast, the work done by scholars points to numerous identities that also intersect with national, regional, and local influences outside of the United States.

The counterclaim to national hegemony was also contained in Gutiérrez and Padilla's statement explaining the research project. "Our mission and goal is nothing less than to recover the Hispanic literary heritage of the United States, to document its regional and national diversity, to view from various perspectives and angles the matrix of power in which it was created, and to celebrate its hybridity, its intertextuality and its polyvocality."[26] That diversity would point toward differences that moved in other directions, away from the U.S. nation. The gesture toward intertextual connections and polyvocal productions paved the way for considering how certain writers and print culture formations did not respect national boundaries. In actuality, some of this Hispanic textual heritage would point simultaneously to two or more sites (Mexico, Cuba, and other countries), bringing forward transnational print culture formations and traveling writers who did not respect national boundaries. In addition, efforts to associate the texts from the past with post-1960s archival claims to liberation projects came under scrutiny as scholars unearthed complicated political alliances and positions in other centuries, most evident in the debates over Maria Amparo Ruiz de Burton, whose position in Chicano literary history clashes with her flirtations with the Confederacy.

Migrant archives began to emerge in the differences of the past. Experiences, writings, and contexts from other centuries did not always fit the paradigm, and not only because these writers may have invented other national or hemispheric American affiliations, but also because they participated in intellectual traditions and political alliances that could not be classified within the dynamics of contemporary Hispanic cultural and political dimensions. One additional migrant element: Newspapers often printed anonymous articles, making it difficult to connect a piece of writing to an author, complicating the very idea of Hispanic literary heritage, and calling attention to the ease with which Hispanic could become Hispanophone U.S. literary heritage. Rather than see the emphasis on language differences as a disciplinary boundary, many of the scholars associated with the Recovery Project took the opportunity to seek out many avenues of research. We are still moving, I would argue, into migrant archives that will ultimately displace the subjectivity that sustains the project.

What emerged was a tension between the necessary evil of the archive, the building that houses the materials, and the challenge of migrant archives. The Recovery Project gathered authority from something that is commodifiable and commonly commodified: the Hispanic subject of the United States as conceived after the 1960s. Would it have been possible to bring the materials of the Recovery Project into national circulation without this archive? Could the documents have staked a claim to truth without the edifice of the project, which has been supported by prestigious foundations and the influence in U.S. society of a growing Latino population? Like the nation and the national archive, an ethnicity and an ethnic archive validate and sustain each other.[27]

Although the phrase "U.S. Hispanic Literary Heritage" claims an association with national literature, it was clear from the beginning that we would need to move outside the boundaries of the nation-state to do recovery work. The project held that door open from the beginning because both temporally and spatially, the Hispanic heritage of the United States has connections to other countries. Those connections sometimes are evident in the content of a text, but also in the print culture conditions or even in the biographies of writers. The Recovery Project has always moved in an interdisciplinary and even indeterminable field, into and beyond American studies, Latin American studies, Spanish-language literature, American literature, and literature of the Americas. In other words, an ethnic subject is a heuristic for a textual reality that is much more complicated. The "Hispanic" in Hispanic literary heritage is like the museum on the Mall version of the National Archives: an edifice that stands in for the intricacies of the many archives held in other locations. The museum is not necessarily an impediment. Actually, it could be a door into migrant archives.

Migrant archives are not widely available, nor is their existence known by a large number of people. They are sometimes in someone's garage or held by a descendant of the person who produced the document. These documents contain stories, experiences, and languages that are not part of easily recognized narratives of institutions. They break out of standard language and official stories. But that is not to say they are completely outside of the physical archive. They move in and out of repositories of rare documents and other libraries. The Beinecke contains migrant archives. So does the Library of Congress. At the Library Company of Philadelphia, founded by Benjamin Franklin in 1731, a collection of German-language documents from the colonial and early republican period is part of a migrant collection. It is migrant because it complicates the common understanding of Philadelphia and of the corpus of myths and stories associated with Benjamin Franklin. It also moves a conception of the early United States outside of the local geography and into a dialogue with the Germanic antecedents of the local print culture.

The documents of migrant archives do not need to be discovered; discovery implies that they are not known and located. Some documents are already catalogued and part of official collections, even though migrant archives do not have their own catalogues. The point is that their presence might not be readily apparent within the existing discourses of academic or political inquiry. In that sense, they differ from new documents by or about well-known historical figures. In the 1980s, for example, a stash of Melville family documents was found in a barn and became part of the lore of research on Herman Melville. Unlike the stuff of migrant archives, these documents had a home in existing academic fields, Melville studies and, more generally, American literature.[28] While fields of study provide insight into the interests and commitments of those who practice within them, who defend them eagerly and/or viciously, migrant archives can contain the writings and visual cultures of those who are dismissed and overlooked by the keepers of the archive. In some cases, the writers are literally migrants. But they might also be members of elite groups who travel first class but whose texts also open a variety of views of the past. Sometimes the routes of migrant archives will lead to new understandings of who and what is excluded from the archive.

The physical archive (the national archive, the official archive) provides many opportunities. As a rare institution, an archive stipulates who can enter. Some archives require registration; others ask for letters of reference. To locate the dusty texts of migrant archives you might have to pass through the door and sign yourself in as a reader at the archive run by the city or the state. The old archive persists. Derrida's injunction against the archive per se calls for a necessary move away from something that might be unavoidable. (The legal problems that arose in relation to the "Derrida archive" are a reminder of the archive's presence.)[29] Furthermore, as the case of the Hispanic literary heritage project shows, some groups do not have the luxury of ignoring or dismissing the importance of archives, nor would it be politically astute to do so.

Migrant archives are calling for the work of those who would go into them and do the careful reading and contextualization that the finest research requires. As pressure builds on national or ethnic archives to account for transnational, exile, and diasporic influences, scholars will have to follow migrant routes rather than revert to the inclusion of a text into a preexisting model. That work is not the primary purview of relatives and descendants of the producers from other decades and centuries, but of scholars who seek to fan the flames of the past so as to create light for the present. Because migrant archives do not have buildings devoted to them, it is up to committed Americanists to locate their contents, read them carefully, and provide contexts for their emergence.

Notes

1. International Council on Archives, http://www.ica.org/.
2. In this article I discuss Jacques Derrida, *Archive Fever: A Freudian Impression*, trans. by Eric Prenowitz (Chicago: University of Chicago Press, 1995). For an overview of debates about the archive, see Marlene Manoff, "Theories of the Archive from Across the Disciplines," *Portal: Libraries and the Academy* 4 (2004), 9–25.
3. Thomas Osborne has argued that the physical reality of an archive, in tension with the abstract conception of what it implies, grants credibility to certain archival configurations and even disciplines. See Osborne, "The Ordinariness of the Archive," *History of the Human Sciences* 12 (May 1999), 51–55. That physical credibility is one factor behind what some have called an archival turn in literary and cultural studies. While historicists' methodologies and the de-centering of the literary text have also contributed to this archival turn, the lack of a consensus as to what constitutes literary studies has led to a privileging of the recovery of texts and contexts that have not been considered in the recent past. Jane Gallop has criticized this historical turn by establishing a dichotomy between close reading and "archival work." Actually, the best historical work would deploy close reading techniques on documents that might otherwise be taken as factual evidence for historical narratives. See Gallop, "The Historicization of Literary Studies and the Fate of Close Reading," *Profession 2007* (New York: MLA, 2007), 181–186.
4. The National Archives also has connections to regional archives and presidential libraries. The United States offers a curious case of archive construction. Unlike other nations, the United States did not establish the institution called the National Archives until more than a century after the country's founding. Individual government agencies had been in charge of their own records before Congress created the National Archives and Records Administration in 1934 (http://www.archives.gov/about/history/). I would argue that the first U.S. "national archive" was the Library of Congress, founded in 1800 when the capital was transferred from Philadelphia to Washington, D.C. Originally intended to house books for use by the government, the Library of Congress became a national library that was to provide information on all subjects, meaning all subjects that would be of interest to the nation. For an account of the founding of the Library of Congress, see John Y. Cole, *Jefferson's Legacy: A Brief History of the Library of Congress* (Washington: Library of Congress, 1993).
5. National Archives Experience. http://www.digitalvaults.org/.
6. Thomas H. Kreneck, Preface to *Crossing the Rio Grande: An Immigrant's Life in the 1880s* (College Station: Texas A&M University Press, 2007), ix–xii.
7. Ibid., x.
8. Guadalupe Valdez Jr., "Memories of My Grandfather: Luis G. Gómez," in *Crossing the Rio Grande*, 14.
9. Derrida, *Archive Fever*, 2.

10. Manoff, "Theories of the Archive," 10.
11. Marc Shell and Werner Sollors, *The Multilingual Anthology of American Literature* (New York: NYU Press, 2000). See also Sollors, ed., *Multilingual America: Transnationalism, Ethnicity, and the Languages of American Literature* (New York: NYU Press, 1998).
12. "By this term [archive]," Foucault writes, "I do not mean the sum of all the texts that a culture has kept upon its person as documents attesting to its own past, or as evidence of a continuing identity; nor do I mean the institutions, which, in a given society, make it possible to record and preserve those discourses that one wishes to remember and keep in circulation" (Michel Foucault, *The Archeology of Knowledge* [New York: Pantheon, 1972], 128–129). Instead, Foucault would focus on the historical discursive system (assumptions, connections, relationships) that permits a particular statement. Libraries and academic disciplines or fields are the product of the archive but are not the archive themselves. If a national archive implies continuity and tradition, Foucault's "archive" functions in a particular context. Rather than focusing on the content of an archive or the person who inaugurates or promotes it, Foucault seeks to understand how the archive delimits what and how something can be said.
13. Derrida, *Archive Fever*, 45.
14. Derrida, *Archive Fever*, 20.
15. For a discussion of "radical evil" in various strains of philosophy, see Richard J. Bernstein, *Radical Evil: A Philosophical Investigation* (Cambridge, UK: Polity, 2002). For a brief explanation of Kant's concepts in relation to other uses of "evil," see Christoph Cox, "On Evil: An Interview of Alenka Zupancic," *Cabinet 5* (Winter 2001–2002).
16. Carolyn Steedman, *Dust: The Archive and Cultural History* (New Jersey: Rutgers University Press, 2002), 27.
17. Steedman, *Dust*, 27.
18. Lawrence Venuti, *The Translator's Invisibility: A History of Translation* (New York: Routledge, 1995), 7.
19. Kirsten Silva Gruesz calls for overcoming a "lack of interest in questions of language difference and translation in most Americanist work at present." See "Translation: A Key(word) into the Language of America(nists)," *American Literary History* 16 (2004): 90. In developing an approach toward hemispheric studies, Susan Gillman argues that *adaptation* rather than translation may be a more fruitful way to think about texts that are incommensurable and affected not only by different literary histories but also uneven economic and social conditions in the Americas. See Gillman, "Otra Vez Caliban/Encore Caliban: Adaptation, Translation, Americas Studies," *American Literary History* (Spring/Summer 2008): 187–209.
20. Derrida, *Archive Fever*, 4f.
21. Jacques Derrida, *Mal d'Archive: Une Impression Freudienne* (Paris: Galilée, 1995), 15f.

22. The project has also awarded many fellowships for research and sponsored a biannual conference that has become a central gathering place for many scholars interested in the Hispanic past. The conferences, in turn, have led to a series of volumes of critical articles.
23. Ramón A. Gutiérrez and Genaro M. Padilla, "Introduction", to *Recovering the U.S. Hispanic Literary Heritage* (Houston: Arte Público Press, 1993), 21.
24. Walt Whitman, "The Spanish Element in Our Nationality," *Poetry and Prose* (New York: Library of America, 1996), 1146.
25. Ibid., 1147.
26. Gutiérrez and Padilla, "Introduction", 21.
27. A similar process is evident in the racial authorization of African American studies, although not without its own debates. As Xiomara Santamarina has noted, a methodological paradox emerges in that field when scholars develop an archive presumably about a racial group at the same time that certain practitioners deconstruct the racial particularity that propels such a search. "[T]he field encompasses reprinting and reinterpreting long out-of-print and forgotten texts, in combination with textual and contextual analyses that recover the instability and contingency of racial discourses across space and time." In response, Santamarina argues for the ongoing recuperation of archival materials that call attention to and speak to the need for specific analyses—in this case, historicizing practices that place discourses on race in the particular conditions faced by the writers. But one has to wonder whether in African American studies race remains as the category that sustains the archival process. By contrast, migrant archives move away from racial formations. See Santamarina, "Are We There Yet?": Archives, History, and Specificity in African-American Literary Studies," *American Literary History* 20 (Spring/Summer 2008): 304.
28. Hershel Parker opens the first volume of his monumental biography of Melville by emphasizing the importance of that discovery. See *Herman Melville: A Biography, Vol. 1* (Baltimore: Johns Hopkins University Press, 1996), xi.
29. The competing claims on Derrida's papers and Derrida's response to institutional injunctions show just how important and even scandalous archives continue to be. For a newspaper account of these troubles, see Thomas Bartlett, "Archive Fever," *The Chronicle of Higher Education*, 20 July 2007, A8.

CHAPTER 11

Partnering Across the Americas: Crossing National and Disciplinary Borders in Archival Development

Melissa Bailar

As the preceding two essays illustrate, collaborations, research and pedagogical tools, and materials that facilitate modes of inquiry spanning national boundaries are of increasingly exigency to those interested in teaching and studying the Americas. Developing these avenues of inquiry is perhaps more pressing than ever before, as violence in border towns has increased, epidemics have caused governments to periodically shut down travel among nations, immigration regulations have tightened, and acts of terrorism have reinforced fears and hostility about otherness. Jorge Bustamante, the United Nations Special Rapporteur on the Human Rights of Migrants, has insisted that when we approach border conflicts as either domestic or foreign policy issues—in other words, along a unilateral or "us-versus-them" perspective—nations reach a state of opposition and ultimately impasse.[1] More productive outcomes can be achieved, he suggests, when we view these issues from a broader and multifaceted—a human, rather than nationalistic—perspective. To change the way in which broad social questions such as immigration or education are investigated, government and institutional structures need to implement multinational perspectives. Yet, as many of the contributors to this volume have observed, knowledge remains largely organized in nation-based categories. Bustamante's comments thus have consequences not just for politicians, but also for the scholars and teachers who help articulate and shape attitudes and approaches towards questions of social concern.

Projects that strive to restructure knowledge categories according to perspectives not based on the concept of nation must work through practical questions about how hemispheric paradigms transform research tools and archival collections. Despite inherent difficulties imposed by differences in language, culture, national regulations, and economic priorities, multinational collaborations that develop such resources are necessary if scholars are to collect and disseminate the richest possible collection of documents and ideas from across the Americas. The Our Americas Archive Partnership (OAAP),[2] a multiinstitutional federated digital archive, provides a compelling example of one such collaboration (see http://oaap.rice.edu and Figure 11.1).

This project strives both to provide an archival product that facilitates hemispheric research and teaching, and to adopt a hemispheric approach in its development and dissemination. The processes of imagining, designing, and implementing this archive have provided concrete examples of the challenges and opportunities that practical applications of hemispheric theories raise.

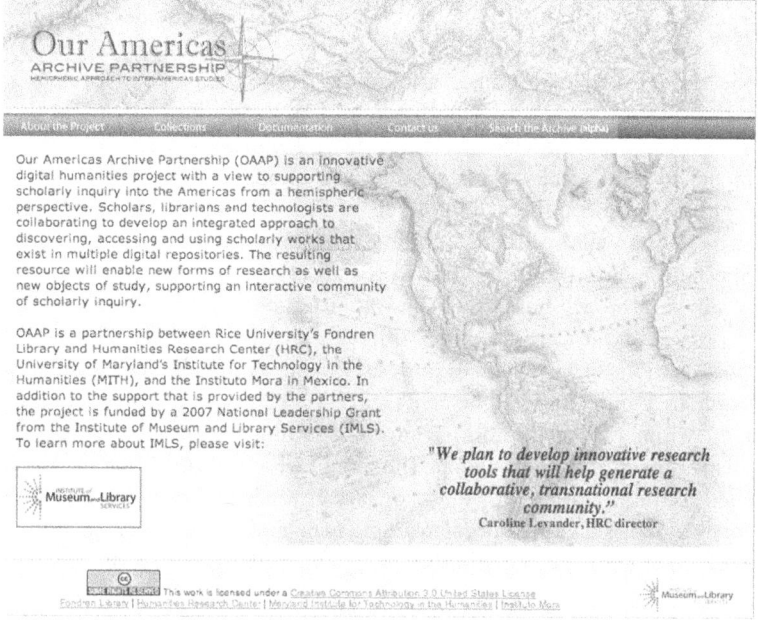

Figure 11.1   The Our Americas Archive Partnership web site

Source: This project, including all images in this essay, is licensed under a Creative Commons attribution 2.5 License.

Scholars committed to taking on hemispheric questions confront serious institutional obstacles. As Bustamante suggests, productive resolutions to problems that stem from the interactions of different groups require strategies that integrate multiple viewpoints, yet the necessity of multinational intellectual collaborations, for instance, has outpaced the structural changes needed to facilitate such partnerships. Resources for such collaborations thus remain scarce. In the United States, many government grant-making programs will not cover salary for researchers based at non-U.S. institutions, the purchase of equipment for use by institutions in other countries (such as scanners to capture page images of documents housed in remote locations),[3] or even airfare on non-U.S. carriers except in the most exceptional of circumstances.[4] Thus, federal grants in the United States tend to reinforce nationalist mentalities. Similarly, historians funded by governments in Latin American and Caribbean nations are often obligated to promote their country's unique history, rather than illustrate how its stories overlap or interact with those of other nations. Documents housed in any nationally funded institution, moreover, are usually not permitted to leave their country of origin without extensive and costly paperwork and precautions. Thus, while some politicians, scholars, and other individuals recognize the urgency of adopting hemispheric approaches, current infrastructures frequently thwart efforts to do so. Without tangible examples of the feasibility and utility of such partnerships, policies are unlikely to change. Designing and instituting small hemispherically oriented collaborations is necessary to begin to pave the way for future large-scale projects that do not just illustrate but fully incorporate an outlook that traverses many borders. As a kind of pilot for future transnational archival projects, the OAAP has confronted several practical obstacles that are instructive for the future development of many new multinational and cross-disciplinary collaborations.

Three institutions, Rice University in Houston, Texas, the University of Maryland's Maryland Institute for Technology in the Humanities (MITH), and Instituto Mora in Mexico City (with future partners anticipated), have collaborated in the OAAP's development, though each has entered this project from a different starting point. The University of Maryland's Early American Digital Archive (EADA), one of the first collections to span national and disciplinary borders in the Americas, compiles digital versions of texts originally written between 1492 and 1820.[5] EADA grew out of the 2002 Early Ibero/Anglo Americanist Summit, in which 100 scholars in different fields exchanged ideas about hemispheric approaches to researching and teaching the Americas, and stressed the need for an efficiently accessible and economically feasible collection of texts spanning the Americas. The online anthology of texts that participants submitted to serve as the basis for their discussions was enhanced with search functions and grew into the still-expanding EADA. Any scholar may submit text files of original documents that fit the EADA's parameters, with introductions and annotations if desired. Submissions are peer-reviewed and proofed against the originals by an editorial panel before MITH integrates them into the EADA and its search tools. Because EADA compiles digital versions of texts that reside in widespread libraries and private collections, the archive is not attached to any physical repository. One of EADA's primary concerns is that it remain free of charge, and so it does not support the often-expensive and labor-intensive scanning and maintenance of a visual collection of the page images of its texts.

In contrast to the purely digital EADA, Rice University purchased many of the documents it contributes to OAAP, with the express purpose of collecting a sample of texts that highlight the interconnectedness of the American hemisphere,[6] and a few items already in Rice's rare-books holdings were added to these. Most date from after 1820 to round out the EADA's offerings. This selection of about 26,000 pages of texts and images, which all physically reside at Rice, are scanned on a high-end color scanner, transcribed, and marked up so that they are searchable online. While it has fewer items than the EADA and has been more expensive to develop, the Rice archive provides more "layers" of its texts, with a physical archive, page images, transcriptions, and some translations.

The third partner, Instituto Mora (http://www.institutomora.edu.mx/), possesses an extensive physical archive of textual, visual, and audio materials on the history and social sciences of the Americas from the eighteenth century through the present but has a limited scanning system and little sustainable server space for digital storage. Further, because Mora is in Mexico and much of the funding for OAAP was granted by IMLS (a U.S. federal institute), their collaboration has been supported predominantly through private donors and volunteer efforts. Ensuring that Mora's participation fits within the grant regulations and limits of available resources has allowed for a little more than 9,000 of their document pages to be scanned on Mora's own equipment in black and white and housed on Rice's server, with hopefully more to come. These documents are then transcribed and digitized in the same ways as the Rice documents. The three collections, then, are distinct in their approach and appearance. The OAAP's technological challenge has been to develop an interface that allows for seamless searches across these three different sets of materials, as well as potential future collections with still different parameters. A user, then, could search for documents using the key word "labor," for example, and the results would identify relevant documents in all collections included in the OAAP.

Documents that complicate the prevailing static notions of nation and border, and emphasize movement and overlap among nations in the Americas have been given priority in the OAAP's selection of documents. Previously neglected documents, typically manuscripts written by little-known authors for personal use, reveal the human side of encounters with otherness and sameness in different regions. For example, George Dunham, an engineer who traveled from Boston to Brazil to design equipment for a sugar plantation, describes in his journal the circulation of contraband copies of *Uncle Tom's Cabin* among the Brazilian population. He writes, "I find that *Uncle Tom's Cabin* has got into Brazil and the people will read it. It is translated into Portuguese by a French man and several of them have got into the country but the Government has prohibited the sale of it. I have seen a Brazilian that can read English reading a book that he appeared very cautious about any one seeing the title of but I saw on the cover, *Uncle Tom's Cabin*" (168–169). The novel's roundabout translation from English through a French intermediary into Portuguese emphasizes the distance between the populations reading it, but its circulation despite obstacles imposed by linguistic and governmental barriers is testament to their shared struggles and emancipatory sentiments. The Dunham journal, then, traces not just the movement of an individual between nations and his ensuing reactions, but also the cross-national journey and impact of a text. Documents such as this capture the historical circulation and cross-influence of ideas among diverse populations, and their inclusion in an archive shaped along a hemispheric perspective fosters the continuation of such exchanges by providing a multinational community of scholars with access to these very documents.

An unusual document in the OAAP illustrates concretely the way in which the circulation of documents across space and time can expose interconnected histories—a piece of paper currency printed in 1823 for use in Stephen F. Austin's colony in Texas. At that time, Mexico had been independent from Spain for two years and Texas was a Mexican territory. The note worth "diez pesos" was printed on the back of a papal bull that had been printed in 1784 and distributed in Mexico, then under Spanish rule (see Figure 11.2).

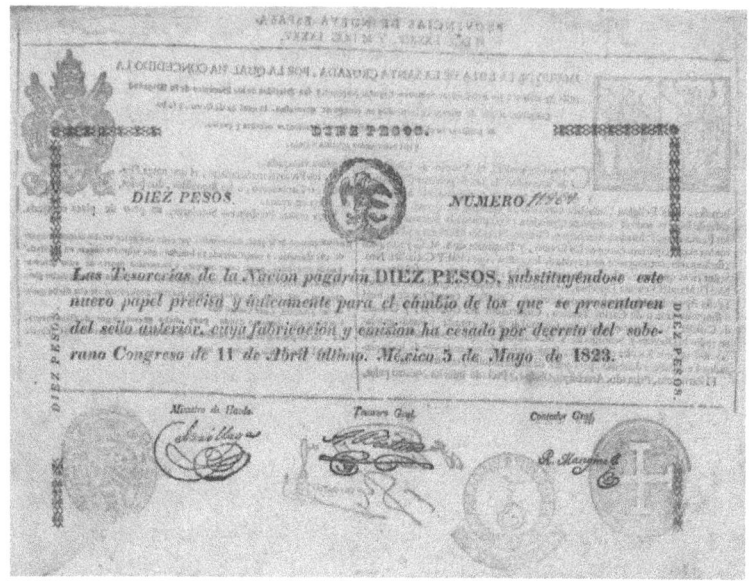

Figure 11.2A This side of the document functions as currency

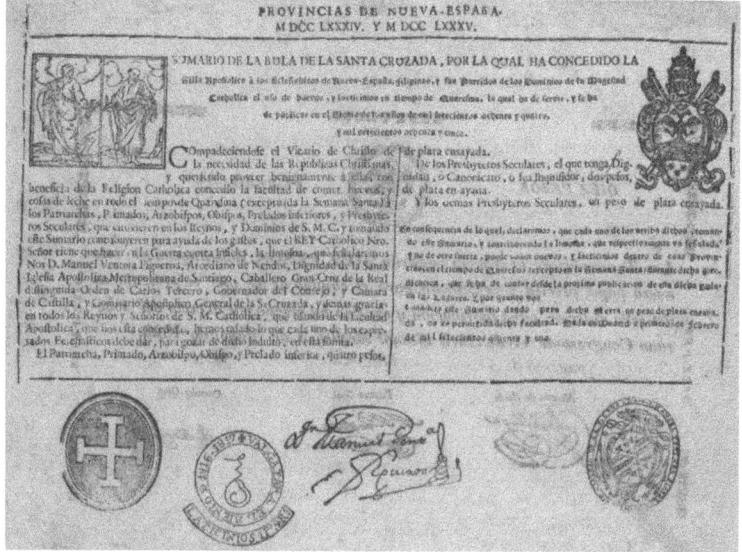

Figure 11.2B This side of the document is a papal bull

Although the front and back of the paper have quite distinct subject matter, intended audiences and uses, and appearances, their physical juxtaposition highlights an aspect of the history of this region not evident in the content of either side: the severe paper shortage in Mexico after the dissolution of the Mexican Empire in 1821. This item has posed challenges in chronological and geographic assignments by archivists who want to capture the three distinct stories of currency, papal decree, and paper shortage this one document relates. In the end, Rice archivists opted to treat this item as two separate documents with their own sets of metadata; this solution, however, neglects the story revealed only through their juxtaposition. Ideally, a hemispheric archive would similarly bring together distinct documents to reveal other stories not readily visible in research that stops at traditional borders.

The convergence of texts from the three collections in the OAAP forms clusters centered on themes such as revolution or emancipation that have affected the hemisphere. The texts in these clusters may have been housed in different locations, written in different languages, and

Figure 11.3 A color plate of an Uruguayan woman in *Las Mujeres*

published in different countries and centuries, but together they reveal overlapping histories that are not limited to any one nation. A book in the Rice collection published in 1876 in Mexico City called *Las Mujeres españolas, portuguesas, y americanas* is a striking collection of thirty chapters written by (male) authors on various aspects of the customs, dress, and responsibilities of women from different countries in the American hemisphere. Each chapter is accompanied by a full-color illustration of a woman in her native traditional dress (see Figure 11.3).

Although filled with national biases, male voices interpreting and masking female experience, and even instances of "plagiarism," this work nonetheless represents an early attempt to collect considerations of one topic from multiple nations. Instituto Mora has also contributed two little-known books published in Spain on national differences in the customs of women, *Historia moral de las mujeres: influencia de la mujer en el progreso y cultura de las naciones* (1889) and *América y sus mujeres* (1890). These three books, along with additional manuscripts detailing women's perceptions of historical events, form a new set of research materials on gender history housed in one digital location.

Page images of diverse documents may also be prepared in ways that make them easily discoverable by interested researchers, teachers, and students. The efficiency with which a scholar can identify items of interest improves when transcriptions of handwritten items are available. Not only does a typed version allow for quick and easy reading to determine if more careful study of a text would be fruitful; it also enables the tagging and coding that make documents come to the surface, with different search tools that look for specific languages, place names, dates, or other information embedded within a text. In any archive that comprises diverse materials in different formats and languages, the documents that are easiest to read, written in a language with which the researcher is familiar, or written by someone well known, are likely to be the most studied. In the Rice Americas archive, approximately 2,000 pages of documents are handwritten items authored by little-known persons that demonstrate an individual's perspective or experience of movement and influence across national borders. So that these would rise to the attention of researchers and teachers interested in hemispheric approaches, and so that easy-to-read printed items would not remain privileged, the OAAP has opted to include transcriptions of each page. The purpose of the transcriptions is to increase the efficiency of the archive, not to replace careful study of the originals.

The process of transcription is not limited to technical concerns, as it in fact highlights several of the methodological challenges imposed by hemispheric studies. Many of the specific challenges facing the transcribers represent on a microlevel the complexities inherent in broad, theoretical hemispheric perspectives. The content and format of documents that illustrate interactions among peoples of different cultures regularly include linguistic and visual cues that the author has experienced something unfamiliar, such as attempts to use words in a different language (which may be spelled phonetically), additions and annotations to clarify descriptions, references to places that have been occupied by different peoples and have accordingly changed names, and sketches or maps to visually represent new information. Thinking through how to transcribe such instances requires the transcriber to reflect on the larger goals and challenges of hemispheric approaches to ensure that pertinent information is not ignored or presented in a misleading or overly simplified way. Often, the items that are most technically difficult to transcribe, such as travel journals, hasty accounts from the frontlines of revolutions, and mixed-language letters written to recipients of different nationalities, illustrate through their content and their use of language the sorts of conflicts and influences of primary interest to hemispheric Americas researchers.

Transcriptions are also central to debates about changing approaches to archival research. With the rapid increase of digitized material available, it can be difficult to assess the validity of different sources and tempting to rely on materials several steps removed from original documents.[7] In the best of circumstances, transcriptions alter or obscure important information—the neatness or hastiness of the handwriting, the condition of the paper, the writing instrument used, the relative proportions of blank spaces, items taped into journals, and nonlinguistic marginalia, to name but a few aspects. In most cases, transcriptions contain errors (a mistranscribed word, the inclusion of a word not in the original, or the omission of a word) that can shift the meaning of phrases, potentially altering the impact of the entire document.[8] To encourage scholars to refer to the original page images, rather than rely solely on transcriptions (performed by students and staff members under pressing deadlines), the Rice and Mora digital archives include at the top of each transcribed page thumbnail images that link to full-page scans of the manuscript.

The concrete difficulties a transcriber faces potentially increase in a hemispherically oriented archive that spans several centuries, languages, and nations; thus, transcriptions are best performed by someone familiar with the themes of the archive and the literature and history of the geohistorical period of the document under consideration. This

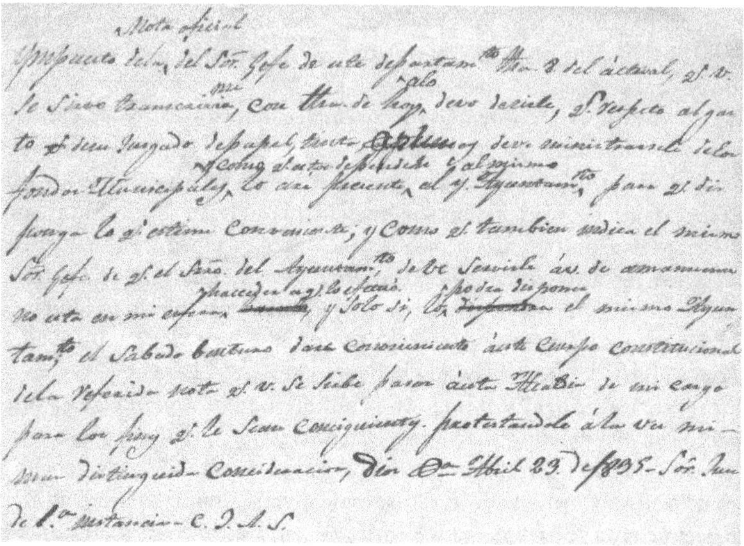

Figure 11.4A This figure shows a scanned section of page two of "Carta e informes con respeto a la administración del gobierno municipal y el tribunal, con una petición por un juicio lenitivo en el caso de José Antonio Delahoz"

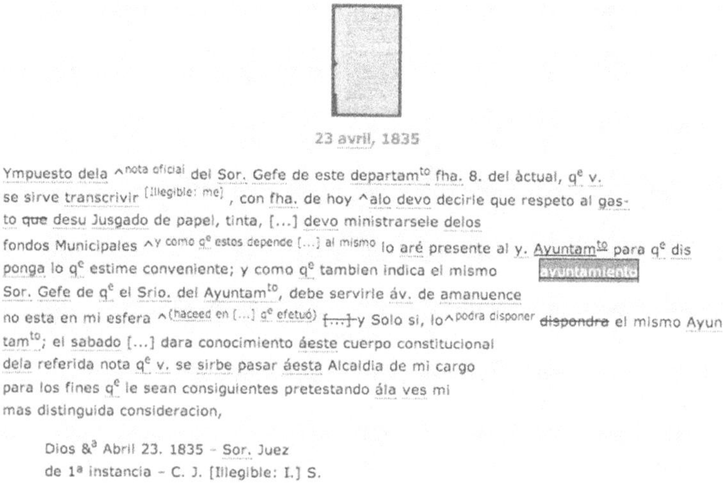

23 avril, 1835

Ympuesto dela ^nota oficial del Sor. Gefe de este departam.^{to} fha. 8. del àctual, q^e v. se sirve transcrivir [Illegible: me] , con fha. de hoy ^alo devo decirle que respeto al gasto que desu Jusgado de papel, tinta, [...] devo ministrarsele delos fondos Municipales ^Y como q^e estos depende [...] al mismo lo aré presente al y. Ayuntam^{to} para q^e dis ponga lo q^e estime conveniente; y como q^e tambien indica el mismo 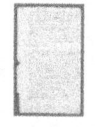 Sor. Gefe de q^e el Srio. del Ayuntam^{to}, debe servirle áv. de amanuence no esta en mi esfera ^(haceed en [...] q^e efetuó) [...] y Solo si, lo^podra disponer dispondra el mismo Ayun tam^{to}; el sabado [...] dara conocimiento áeste cuerpo constitucional dela referida nota q^e v. se sirbe pasar áesta Alcaldia de mi cargo para los fines q^e le sean consiguientes pretestando ála ves mi mas distinguida consideracion,

Dios &^a Abril 23. 1835 - Sor. Juez
de 1ª instancia - C. J. [Illegible: I.] S.

Figure 11.4B This shows a portion of the HTML version of the handwritten document, including regularizations

allows, for example, for the transcriber to recognize terms, names, and places not commonly employed today, such as "Texian" or "Goliad." Documents that discuss multiple nations often contain phrases written in different languages that may have spelling errors if they are not in the author's native language. Ideally, transcribers would have a grasp of all languages they encounter, which would allow them to figure out unclear or misspelled words from their context (and also commit fewer transcription errors). The better transcribers understand both the language used and the content and import of a document, the better they can decipher author-specific shorthand as well. For example, the vague "R¹ Aud^a" might stand for "Real Audiencia," but this only becomes obvious given the subject matter in question.[9] Many documents included in the OAAP contain illegible words, sometimes several on a single page. Some of these words can be deduced given the context, and deduced with a greater rate of accuracy by someone familiar with the subject matter. To encourage scholars to consult the original page images when guesses have been made, guesses are distinguished as such in the OAAP. For example, if a word is only likely to be "William," it is transcribed as: [Illegible: William].

Handwritten documents often contain spelling and grammatical errors, especially diaries and letters written by persons with limited education, wealth, and social standing. Such irregularities provide clues to their authors' lives, but they can also make the document hard to understand. Because the audience of a hemispheric archive would ideally be multinational, such documents, even transcribed, might pose challenges to users without a strong grasp of their original language. Typically, then, transcribers in the OAAP have maintained the irregularities within the documents, but they also provide corrected versions. Thus, someone who is keenly interested in the author or the originality of the document can examine all of its irregularities, while someone interested in garnering an overview of the manuscript's content can read quickly through it, and someone less familiar with the language used in the text can perhaps grasp the sense of the document. (See Figure 11.4, which shows the scanned image of an original handwritten item and its appearance on screen with one regularization highlighted.)

After the document has been transcribed, the next step to making it appear readable on screen is to code it so that it is accessible online in HyperText Markup Language (HTML) format according to the text-encoding initiative (TEI) guidelines, a set of standards for preparing digital documents for scholarly use.[10] For the OAAP, we have opted to employ an external vendor for the HTML conversion of printed documents, since they are typically far more straightforward than manuscript items in terms of both their readability and their overall structure. On the other hand, students, staff, and scholars in library sciences and Americas studies have performed all mark-up of handwritten items. This cross-disciplinary collaboration on the TEI preparation has been invaluable for making decisions about the most user-friendly and comprehensive ways to treat documents in a varied and asymmetrical collection. Such considerations include, for example, determination about what counts as a section "head" (heads generate tables of contents, which risk becoming too detailed and cumbersome, with hundreds of divisions, or too brief, with only superficial divisions to facilitate navigation through a document); how to display regularizations of proper nouns (this allows search engines to bring up original misspellings such as "New Orlins," for example, when someone wants to find all mentions of "New Orleans"); and how to clearly distinguish among notes in documents that include original footnotes, annotations marked by someone other than the author on the document after its original composition, and editorial explanations added by a transcriber or scholar. These are a few of the many decisions that were discussed over many months among people from multiple fields to ensure that while adhering to rigorous library standards the digitization process keeps the user population of scholars and teachers in mind. Although such considerations are technical, they do require an understanding of the theoretical goals of the archive to ensure that documents are prepared properly for the new sorts of searches useful in hemispheric inquiries. None of these considerations is value-free, each requiring decisions on what is important to tag, highlight, annotate, and omit. The approach to TEI differs widely in collections prepared for different audiences and serving different goals.[11]

An additional step towards making this digital archive accessible to a widespread and varied audience, one that has proven to be quite complex, is to make at least portions of it available in different languages. Ideally, a hemispheric archive would be multilaterally translated, equally accessible to people speaking all the languages represented within it. In the OAAP, for example, that would mean that all documents would have an English and a Spanish version, since the archive features these two languages above all others (roughly a 60:40 English to Spanish ratio), but also perhaps French, Portuguese, and Dutch versions, as a few documents are in these languages as well. As the archive grows, additional items in these and other common languages, as well as indigenous languages and local dialects, may also be added. Such a multilateral approach might be ideal, but would be beyond the capacity of individual translators hired onto the project and of scholars who might generously donate time for such efforts.

The criteria for decisions regarding approaches to translation differ according to whether the documents within the archive or the digital tools are considered. Before encountering the documents, users first navigate through an online interface that includes general information on the OAAP and its partners, an array of search tools, and metadata of the documents (metadata includes publication information, Library of Congress subject headings, details on the digitization process, information on the physical location of the items, and more). This information is fundamental to any new visitor to the archive, and making at least the structure and theory behind the archive accessible to a wide audience through multilingual versions is a key step towards securing new partners in the project and fostering future hemispheric endeavors. Because the interface will continue to change as new collections (each with its own set of metadata records) are added, as scholars add search tags or otherwise refine the tools, and as, perhaps, the archive grows and its goals shift, static translations of its components would need to be updated and expanded frequently. Therefore, the only feasible option for a fully multilingual interface appears to be to rely on translation software that can transform the interface into other languages in real time. The OAAP is currently considering keeping the default language of the interface English and incorporating a free application called the "translate widget" from Google Translate[12] into the site. This allows users to click one button to roughly translate the interface and tools into their preferred languages. Although this tool provides some awkward translations and does not consistently translate all the clickable buttons (which may be graphics rather than text), this approach does allow a wider audience greater accessibility to the archive, and hopefully translation applications will continue to improve.

Some of the difficulties in using such a translation tool are specific to the linguistic lack of sophistication of the application (the transformation of "Rice Americas Archive" into "Archivo arroz Américas," for example), though others raise more abstract questions that are beyond the scope of software to resolve. For instance, one of the thematic search terms in English is "liberalism." The direct translation of this word into Spanish, "liberalismo," however, has an additional meaning beyond that in English, for throughout Latin America it primarily refers to the postcolonial movement to separate church and state. A Spanish speaker unfamiliar with the nationally different connotations of this word might be surprised that a search using this term turns up documents relating to U.S. political reforms and very few that mention anticlericalism. There are many such instances of significant meanings being lost or distorted in translation, and recognition of them is a challenge in projects that engage multinational users. At the document level, a translator can include notes to explain the differences in meaning, but at the interface level, this is not possible with an automated tool and without cluttering the interface with annotations. When more than two languages are involved, it becomes even more difficult to capture different meanings and maintain the usability of the site for research rather than for linguistic explorations. The OAAP, therefore, plans to let these irregularities remain as they are for users to resolve, as there is no way to predict all of the linguistic complications across multiple languages that will arise as new partners, materials, and interactive components of the interface are added. At a future point, scholars will be able to add commentary to the site, which may help with the identification and clarification of troublesome translations.

Because such differences in meanings across nations are frequent and of great significance in the study of texts, the OAAP has employed a far more intensive translation strategy at the document level. In order to make as many documents as possible accessible to the most people expected to use the archive[13] within the restrictions of available resources and time, the OAAP includes English translations of portions or the entirety of nearly all of its works that were originally written in Spanish. The OAAP has focused on translating all handwritten items, entire previously untranslated books that seem especially rare and provocative, and portions of most other Spanish-language documents not known to have existing translations, such as some preambles to short-lived constitutions and prefatory materials. As time and funds allow, Spanish translations of key English documents are added. The project hired Lorena Gauthereau-Bryson, a skilled translator with research experience in the history of the hemispheric Americas, to do the majority of the translations.[14]

This decision to translate mostly in one direction, and into English moreover, can easily engender criticism, most notably that it undermines efforts to adopt a perspective that does not privilege the United States and that it reinforces nationalistic and colonialist mentalities. This is a key concern for a project that wishes to facilitate non-nation–based and non-U.S.–centered study of the Americas. This decision diverges from the goals of the archive on a theoretical level, and thinking through the difficulties in translation efforts exemplifies the practical implications and challenges of the theories behind hemispheric studies. Indeed, many small and concrete steps towards the development of this archive have encouraged a pointed and concrete consideration of larger issues facing the hemisphere. Discussions of how and whether to translate documents weave into considerations of foreign language education, public education availability in unofficial languages of a nation, pressures on immigrants to assimilate, and other such issues of immediate social concern.

While the OAAP recognizes that a fully multilingual digital archive would be the approach most faithful to the archive's theoretical goals, practical considerations of grant restrictions, available funding, and the personnel hours needed for a multidirectional translation effort have made it thus far unfeasible. The OAAP had to decide whether to include predominantly unidirectional translations or none at all. Because the majority of the people working on the project and the expected user base in its first years are predominantly from the United States (efforts are underway to expand and diversify both of these groups), providing English versions of Spanish documents is a first step to making the contents accessible to a wide population of teachers and scholars. By making Spanish-language documents comprehensible to those who understand English (and who may be native speakers of neither language), the OAAP hopes to begin to break down one barrier to a broad hemispheric mentality. The Spanish-to-English translation process is, in many ways, a test case to explore the possibilities of and obstacles to larger translation projects, and has engendered fruitful discussions about how to apply broad theoretical goals to practical considerations.

The translations of items in the OAAP are not treated as stand-alone items, for it is not the translation that we wish to be studied; rather, the translations provide access to rare documents that are themselves the objects of interest. Just as the project's transcriptions are intended not to substitute for but rather to provide online users with greater accessibility to the page images of the original documents (where available), the translations are considered as supplements to the original works and are in no way meant to replace the Spanish-language originals. Taking this chain of nonsubstitution one step further, archivists have repeatedly noticed that when they make their rare collections digitally available, in-person visits to the physical archives actually increase.[15] It seems, then, that generally no alteration of the actual document is understood by the public to replace a "truer" version of it. In each case, the versions of a document that are removed from the original exist to provide for more efficient perusal of the document, but are not intended to diminish the value of study of the original.

The process of translation has revealed instances of linguistic complexities and multinational influences that otherwise might have gone undetected, and more such discoveries will undoubtedly occur with further translation efforts. One such example arose with the aforementioned *Las Mujeres españolas, portuguesas, y americanas.* Each chapter of this book focuses on a different location (with varying degrees of emphasis on the women of these regions) and was written by a different (male) author. In researching irregularities in Nicolas Ampuero's chapter on Ecuador, the translator, Gauthereau-Bryson, discovered that a large portion of the description of the Ecuadorian landscape in this text was in fact remarkably close to the Columbian explorer Francisco José de Caldas's Spanish-language text, "Viaje de Quito a Popayan." Portions of this Spanish text had been translated into French by the French explorer and diplomat Gaspard-Théodore Mollien (with attribution to Caldas) in his "Voyage dans la République de Colombia, en 1823." Ampuero, rather than simply copying Caldas's original Spanish descriptions of Ecuador, to which he perhaps did not have access, appears to have translated Mollien's French version of Caldas back into Spanish, for Ampuero included some of Mollien's summary adaptations of lengthy passages in Caldas and otherwise varied the text from its original form. Ampuero does not attribute these passages to anyone else or make reference to any work containing them. This instance of cross-linguistic "plagiarism" raises layers of questions about the significance of the variations a text undergoes throughout the translation process (in this case, from Spanish to French back to Spanish and then by the OAAP to English) and the impact on different readerships of the linguistic elements within a document. This passage of Caldas's text traveled widely, and at least at the linguistic level, carries traces of its contact with different languages. Had the translator not been researching the historical spellings of city names within this section, this passage's unusual genesis that reveals subtle differences in national perceptions of Ecuador might have gone unnoticed.

An archive project may consider developing new sorts of textual, geospatial, and chronological tools to assist those working on the sorts of questions that implicate the hemisphere and that inform today's social issues with efficient access to the materials they need. This step has proven to be particularly stimulating in the OAAP, for it requires a thorough assessment of the utility and limitations of existing tools within a hemispheric context. One component that sparked intense debate among scholars, librarians, and programmers working on the OAAP was a geospatial searching option. Such tools are becoming increasingly popular and even expected, and Google offers free and easy ways to populate various types of maps with teardrop shapes indicating locations assigned to texts. Close consideration of how such a tool would look and function became increasingly complicated within a context that wished not to emphasize static national boundaries. A standard map of nations with teardrops indicating texts clustered in major cities would in fact obscure two of the most exciting aspects of the archive: the physical movement of its texts across national boundaries and the cross-national exchange and development of information and ideas its texts highlight.

Scholars, archivists, and programmers worked together to think through possible adaptations of such a tool that could allow for a visual and spatial way to discover documents while adhering to the theoretical goals behind the archive. Since political borders have changed frequently and drastically during the 500 years the OAAP addresses, it would seem that the search tool would either have to incorporate this fluctuation or ignore political borders all together. One could conceivably include shifting borders by overlaying maps indicating different national boundaries through time. This, too, poses tricky problems: Would one employ archival maps, many with national biases reflected in the relative size and positions of countries, to determine borders, and if so, how would they need to be altered to overlay one on top of the other? When one is considering the whole of the hemisphere, there are hundreds of dates when borders changed: Would every shift be included and, if so, how would these changes be tracked on maps? How would a contested border or region be portrayed? This multilayered question of how best to represent borders in a historically accurate way that represents the full array of differing national perceptions seems in itself to require a separate long-term project. While beyond the scope of the OAAP grant, scholars agree that a nuanced and dynamic spatial tool that captures the shifts in borders would have a significant impact on teaching and illustrating the fluid place of individual texts in time and space.[16]

The project therefore opted to use a borderless landmass map for this search function, and while this approach avoids the thorniness of political borders, it still faces the question of where to situate texts. The most straightforward approach would seem to be to assign each text to its city of publication, but this information is not always available and this strategy would again efface the inherent movement of texts that the archive wishes to spotlight. The George Dunham travel journal, for example, demonstrates the shared histories and struggles of the Americas and the movement of people, products, and ideas throughout the hemisphere. Assigning such a work to a single point on a map flattens this dynamism. Further, as a personal journal, it does not have a place of publication, and while it originated in Boston and now resides in Texas, Dunham penned its majority in and about Brazil. Even with published texts, the issue is not clear-cut, for Mexico City has often been listed as the default publication place for works written in countries throughout South America. Moving beyond assignments by publication place to a more intuitive interpretation of the main geographical focus of a text, however, requires a subjectivity that could well vary from reader to reader. Further, many texts in the collection, such as the *Las Mujeres* book, treat multiple nations with equal emphasis. The single-location approach to mapping the texts thus is not feasible in an archival project that strives to highlight various crossings of borders.

The OAAP has not yet arrived at a satisfactory solution to this question of a hemispherically oriented geospatial search tool. At present, the tool populates a landmass map with place names mentioned anywhere in the texts. Each teardrop is given the same weight, whether it is a place mentioned in the body of a text or the city of publication. Because an automated program combs the documents for any term that appears in the Getty index of place names, however, the map is overpopulated or mispopulated by passing mentions of places, words that refer to both locations and people or objects, and place names that refer to multiple locations. Furthermore, mention of a country defaults on the map to that nation's current capital. While these are considerable drawbacks to such an approach, this solution does at least visually suggest the cross-national content and impact of documents in the archive. A more comprehensive and nuanced tool that might include both geographical and chronological elements would require careful preparation by a team of scholars, geographic information systems specialists, archivists, and programmers. Consideration of a tool that is quite simple to incorporate into traditional archival frameworks has revealed some of the complexities in restructuring tools to assist with research in a hemispheric context as well as biases inherent in existing research structures.

A more satisfactory attempt to reorganize archival materials has been the assignment of key terms to documents to enable thematic searches centered on clusters of questions that together inform the histories and perceptions of vital social issues. The scholars who initially developed the idea of the OAAP[17] composed a list of rather capacious themes that are of current and historical importance across multiple nations. The terms are: Technical Borders, Colonialism, Conquest, Conservatism, Constitutionalism, Development, Diplomacy, Emancipation, Federalism, Gender, Historiography, Indigenity, Inter-American relations, Labor, Liberalism, Libertarianism, Migration, Nationalism, Neoconstitutionalism, Race relations, Religion, Revolution, Slavery, Technology, and War. A team of researchers and graduate students at Mora, MITH, and Rice assigns these terms, which appear in a tag cloud on the OAAP's searching interface. A user could click on the word "gender" and find the three books on women mentioned earlier, as well as a diary written by a woman fleeing from the Alamo, a note written by Phillis Wheatley's owner to a publisher praising her writing, a letter written by Empress Carlota of Mexico on how local newspapers portrayed her and Maximilian, and an 1878 letter by the social worker and nurse Kezia DePelchin discussing the ostracism and mistreatment of volunteer female nurses in Memphis during the yellow fever outbreak, to name only a few examples. Of course, most items are assigned multiple terms, and searches can be refined by choosing more than one term or by entering criteria into other search tools in the interface.

The OAAP allows users to search for materials using either the standard Library of Congress subject headings (LCSHs) or key terms (or both) to increase the chances of finding rare documents of interest. LCSH categories are far more numerous and specific than the OAAP's key terms, but they demonstrate a strong preference for labeling an item with one physical location. The Dunham journal, for example, has the subject headings "Dunham, George F.," "Travel," "Brazil," and "Description and travel." The journal's key terms, in contrast, are "Emancipation," "Development," "Slavery," "Labor," "Technology," "Race relations," "Historiography," and "Inter-American relations." Together with other refinements, these search options enable a variety of ways to sort through texts according to different needs and preferences. Assigning appropriate thematic terms requires a thorough reading of the text in order to determine all relevant themes addressed within it, and some readers might disagree with the assignments. Later developments of the OAAP will allow for registered users to assign their own descriptive labels to documents and share them with other scholars from different nations and disciplines. This collaboratively built set of descriptions will allow for the shape of the archive and its digital tools to change as research advances.

To bring the attention of scholars and teachers to the OAAP, faculty and students working on different aspects of its preparation have written about ways in which it might facilitate a more hemispheric approach to teaching and research. These essays are integrated into the archive by the Connexions platform, an online environment for sharing scholarly information freely that is growing in popularity, with an international population of teachers. This resource allows users to post "modules" (online essays with visual aspects and hyperlinks) that can then be collected into "lenses" that focus on one theme.[18] Some archival documents link directly to modules that offer suggestions on how they can be integrated into traditional U.S. history and literature curricula (with plans to include examples geared towards Mexican classrooms) or that explain unusual aspects of their history or preparation, and in turn the modules link directly to the documents under discussion. Connexions encourages a collaborative exchange of information so that users may contact authors, annotate modules, and collect diverse materials into new lenses. Some modules engendered exchanges regarding the OAAP within days of their posting, so Connexions should prove effective in circulating knowledge about the archive and hemispheric studies to a broad user base.

Many students and staff working on this project in different capacities have done background research on specific texts to assist with transcription, proofing, translation, or other efforts and have written modules on their findings, whether to explain the unusual history of a rare manuscript, to suggest curricular integration of sets of documents, or to outline the archive's role in emerging digital humanities initiatives. Lorena Gauthereau-Bryson, who discovered the parallels between the Ecuador chapter of *Las Mujeres españolas, portuguesas, y americanas* and Caldas and Mollien's accounts of Quito, for example, has found in the course of her translation work other texts with passages that were copied from elsewhere without attribution. She is now researching the history and legal ramifications of plagiarism in the nineteenth century in different nations and is preparing a module on this topic that will link to the metadata for all documents in which she has found such instances. A select group of advanced graduate students from different fields who adopt hemispheric approaches in their research are composing a set of more than twenty-five modules that demonstrate how specific documents in the OAAP can be integrated into standard U.S. high school curricula to facilitate a hemispheric consideration of a novel, theme, or event. One student, Cory Ledoux, has given a series of talks to high school teachers of advanced placement courses in English and U.S. history, and directed them towards Connexions, both to read the

OAAP team's suggestions for the archive and, should they wish, to post their own modules for dissemination.

One example of Connexions' role in efforts to disseminate the OAAP's broad goals through examination of a specific document is a module by Ledoux that addresses an 1893 letter written by Cuban revolutionary José Martí. Martí is best known for his essay "Our America," on which the OAAP based its name, which has been widely studied and understood to emphasize the importance of the historical and cultural connections spanning the hemisphere and undermining prevailing assumptions that the United States is synonymous with "America." His letter has a more concrete purpose, though it remains just as cross-national and revolutionary in sentiment. Martí advocates for funding of the revolutionary newspaper *El Yara* written by Cuban exiles in Key West, Florida, insisting that it is essential to furthering the movement in which he believes as "the only Cuban paper which, fueled with nothing more than the fervent patriotism of its editor, provides emigrants with the voice of unity and Cuba with the voice of the revolution every day." The Connexions module provides a brief background on the letter, Martí's life, the history of Cuban exile publications, and the political concerns of Cuba in the early 1890s, as well as a bibliography of suggested further reading. Martí himself lived much of his life in the United States while promoting Cuban independence from Spain, and his philosophies grew in part from his extended involvement with a community of exiles and émigrés. This letter suggests that his identity, like that of all Americans broadly understood, is shaped by influences that cross national borders, and it thus encapsulates the theoretical goals shaping hemispheric studies. While this letter is more limited in scope than "Our America," it provides an example of the practical application of Martí's philosophy, and the module suggests that this easy-to-grasp document illustrates the interconnectedness of ideas and movements in different countries so key to a hemispheric mentality. This module and others on slavery, travel narratives, border disputes between the United States and Mexico, and other topics provide examples of documents in the OAAP that can be integrated into students' emerging perceptions of the world as they develop into a new generation of scholars, politicians, educators, and social workers who will confront both the same and new urgent social issues best examined from a multilateral perspective. Although the modules are narrow in focus and suggest specific texts that can be studied alongside canonical works (like the Dunham journal that shows the widespread interest in *Uncle Tom's Cabin*), they point to ways in which teachers can introduce a multinational perspective on discrete topics that may help foster hemispheric outlooks toward broader questions.

Discovering and confronting the varied obstacles to the development of the OAAP has required working collaboratively across nations, institutions, languages, and disciplines. Developing possible solutions to these obstacles has not just resulted in a new set of archival materials and tools for others to explore; it has itself fostered in an academic environment the sorts of multilateral interactions that Jorge Bustamante has called for in the political arena. Projects that require communication across different sorts of borders in a collaborative, rather than confrontational or hierarchical, manner lay the groundwork for the types of multinational and multidisciplinary partnerships needed to change and expand the predominant nation-based framework of knowledge. Such endeavors necessarily mirror in their development the theoretical approaches that their outcomes are intended to promote, and their very undertaking is a step towards a more hemispheric America.

Notes

1. Bustamante presented these perspectives at the Rice University campus in "Trafficking and Vulnerability of Migrants: A Conceptual Form," part of the lecture series *Mexicans Look at Mexico*, November 29, 2007.
2. At the time I am writing this, the archive project is currently halfway through its projected three-year development. Some of the decisions discussed in the following pages may be subsequently altered. Please visit the archive at http://oaap.rice.edu to see its latest version or to contact the project's coordinator with questions.
3. IMLS specifies, "Title to equipment purchased or fabricated with IMLS funds shall be vested in the grantee organization with the understanding that the equipment will be used for the project for which it was obtained." Since grantees are only U.S.-based persons and institutions, equipment cannot be purchased for non-U.S. institutions. *General Terms and Conditions for Discretionary Grants*, p. 10, found at http://www.imls.gov/recipients/administration.shtm. NEH includes similar language and further stipulates, "In expending NEH award funds for equipment and products, recipients and subrecipients will comply with the Buy American Act" (see http://www.neh.gov/manage/gtcao.html#equipment).
4. See, for example, the guidelines for an IMLS grant (emphasis in the original): "All air transportation of persons or property that is paid in whole or in part with IMLS funds must be performed on a U.S. flag air carrier or under a code sharing arrangement with a U.S. flag carrier" (*General Terms and Conditions for Discretionary Grants*, 11). It further stipulates, "The fact that comparable service provided by a foreign-flag carrier is less expensive,

more convenient, or can be paid for with excess foreign currency is not sufficient justification for using a foreign flag carrier that does not have a code-sharing arrangement with a U.S. flag air carrier" (22). See National Endowment for the Humanities: Grant Management, http://www.neh.gov/manage/gtcao.html#foreign for a list of exceptions to this rule.
5. See http://www.mith2.umd.edu/eada to learn more about EADA's genesis and structure and to access its materials.
6. This was made possible by the generous support of the Gilder Foundation.
7. For a discussion of the need to trust online transcriptions of texts and the dangers of doing so, see James Sidbury, "Plausible Stories and Varnished Truths," *The William and Mary Quarterly* 59.1 (2002): 179–184. Many thanks to Alexander Byrd for directing me to this and other essays on transcriptions of the Denmark Vesey court documents.
8. See, for example, Michael Johnson's discussion of the significance of errors in the transcriptions of the court records of the Denmark Vesey "trials" in Michael P. Johnson, "Denmark Vesey and His Co-Conspirators," *The William and Mary Quarterly* 58.4 (2001): 915–976. For a detailed discussion of the difficulties of performing and evaluating transcriptions, see William Proctor Williams and Craig S. Abbott, *An Introduction to Bibliographic and Textual Studies*, 4th ed. (New York: The Modern Language Association of America, 2009).
9. Lily McKeage has worked on the examples provided for the transcription and HTML preparation of OAAP materials.
10. For more information on TEI, see http://www.tei-c.org.
11. The Walt Whitman archive, for example, includes only structural elements, such as line breaks and titles, to avoid any controversial decisions pertaining to thematic elements. The Orlando project takes the other extreme, tagging semantic and thematic elements, and including lengthy editorial notes. Their project includes tags indicating the genre of the text, the other literary activities of the document's author, details about characters mentioned within the text, motives for actions within novels, and much more. Lisa Spiro, director of the Center for Digital Scholarship at Rice University, pointed out these examples and held multiple training sessions for all Rice members working on the TEI for the OAAP.
12. See Google.com (http://translate.google.com).
13. The user base of the OAAP will, hopefully, expand and evolve as the archive grows and includes new partners. The translation decisions will likely need to be revisited in the near future.
14. Lorena Gauthereau-Bryson holds a master's degree in Hispanic studies from Rice University. She was selected to be the primary translator for the OAAP based on her bilingualism in Spanish and English, reading knowledge of other languages, experience in translation, and her graduate-level study of the complexities and dangers of translation efforts. She is working

full-time for two and one-half years translating items for the OAAP. The examples that follow are based on her work on this project and are used with her permission. Other professional translators, who were selected based on samples of their work, translated a handful of other items before Gauthereau-Bryson was selected. She has also consulted with library staff in the translation of passages appearing in Tagalog, Chavacano, and other languages with which she is not familiar.
15. This "common knowledge" among archivists was confirmed through IMLS's National Study on the Use of Libraries, Museums and the Internet, conducted in 2006. See the study's report at http://interconnectionsreport.org.
16. This thought has been expressed by the external board of the OAAP as well as a group of scholars actively engaged in shaping the field of hemispheric Americas who convened at Rice University in fall 2008.
17. Caroline Levander of Rice University and Ralph Bauer of University of Maryland.
18. The general site for Connexions is http://cnx.org, and the lens for hemispheric Americas studies can be found at http://cnx.org/lenses/oaap.

CHAPTER 12

Ghosts of the American Century: The Intellectual, Programmatic, and Institutional Challenges for Transnational/Hemispheric American Studies

Deborah Cohn and Matthew Pratt Guterl

American Studies—one of the most historically prominent venues for the study of the Americas—is under siege on many fronts. The annual conference, once parodied as a bizarre and quirky mix of pop cultural ephemera, is still dismissed as "anti-American Studies." Leaders of the field have been engaged in a lengthy, introspective, somewhat bloody monologue about its complicity in projects of empire, exceptionalism, and global advantage, with most of these debates referring, almost inevitably, to the reckless use of the term "American," and suggesting, in one way or another, a more self-reflexive, transnational, cosmopolitan, or internationalist deployment of that troublesome name.[1] Finally, American Studies is an easy target in an age of budget crises and proto-corporate initiatives. Conceived in a nation renowned for its anti-intellectualism and its supposedly practical "know-how," the field has no easily summed methodology, produces no immediately or instrumentally significant skills for the workplace, and the object of its study is, as some see it, so blurry as to be beyond mere vagueness and imprecision.

In the face of these trends, Indiana University (IU) recently decided to create from scratch a new undergraduate degree in American Studies. And rather than do it on the cheap, as is so often the case, the institution decided to hire new tenure-track faculty from a wide range of fields and disciplines to bring the idea to life. Indeed, the university set out to frame this as a high-end, intellectually tough degree, an "honors-style" major, for a select number of high-achieving students. There would be no effort, at first, at mass production. As a public institution with 35,000 undergraduates in a state with no long history of progressive politics, IU's ambition here was typically strategic: to mine the middle ground between cutting-edge pedagogy and intellectual practice, stressing higher academic standards as a prophylactic measure against the common charge that interdisciplinary labors are inherently vague and shallow. The local collaborations, compromises, and consequences that accompanied the degree's proposal and approval clarified the specific features of the degree, but they also highlighted the connection between one university's past, present, and future. One does not have to be a futurist to see these details as reflections of the nationally shifting research profile of American Studies as an interdisciplinary and transnational field, nor to recognize that the partnerships forged on the Bloomington campus were evidence of bigger trends across the humanities and within area studies in the United States. In the pages that follow, we want to illuminate what was institutionally unique about the story at IU, and what wasn't.

We write as recent stewards of this program, having served as director and associate director together from 2005 through 2009. With a wonkish attention to institutional specifics and curriculum development, we lay out the specifics of a story we know well. Our hope is not just that the essay offers instructive support for people interested in rebuilding, or "reworlding" American Studies, but also that it illuminates one very local translation of abstract ideas into a more globally minded curriculum, pedagogical philosophy, and hiring practices. This story is, we submit, a window onto a certain kind of becoming, and from which we can see the foundations of a new/nouvelle/nuevo American Studies. If it is a reflection of our sustained focus—in written work, at the ASA conference, and in private dialogues with colleagues inside and outside the United States—on the hemispherics of this broader American Studies, it is also a reminder of the significance of institutional dynamics in enabling, channeling, and limiting our intellectual energies in the day-to-day.[2] And it reveals—as if we needed even more proof—the durable, earthbound challenges to be faced by the ascendant paradigms of the field.[3]

Undergraduate Means

Our story begins in January 2003, when the College of Arts and Sciences welcomed an external review committee to campus, chaired by Janice

Radway, former president of the American Studies Association and a professor in the Program in Literature at Duke University, and including Barry Shank of the Comparative Studies department at The Ohio State University and Thomas Gieryn of IU's Department of Sociology. Noting the long-term commitment of faculty, staff, and resources to a preexistent graduate program, the committee made several recommendations, the most prominent of which concerned the creation of an "intellectually rigorous, honors-type undergraduate major that would place the concept of 'critical citizenship' at its heart," and the establishment of an independent, stand-alone PhD in American Studies as foundational support for these new degrees, the committee recommended that the program be given permanent, tenure-track FTE (Full-time equivalent) faculty, including one senior appointment and an additional staff-person.[4] The result of this "modest investment of resources," their report stressed, would be a "top ten program." However one feels about such rankings, this was a useful, constructive report that set the path that was followed for the next six years.

An unusual financial quirk brought about the funding for this revolution. To facilitate new curricular innovation, the university began collecting a new set of student fees to be used to underwrite a one-time, systemwide competition for funding for new programs and initiatives known as the "Commitment to Excellence" (CTE). Soon after the external review report, a CTE proposal for a "New Focus in American Studies"—written by former Director Eva Cherniavsky, former Dean Kumble Subbaswamy, and others—was awarded $527,000 in new base expenditures to provide for the hiring of four total FTE, one new staff-person, and considerable resources in support of undergraduate teaching and campus programming. Cherniavsky's draft of the proposal emphasized outreach to K–12 teachers and a global focus on democracy studies, a focus that would be in keeping with the university's worldwide reach and impact and the college's faithful commitment to interdisciplinary programs as laboratories of curricular innovation. But by revising the proposal in a brief, unofficial, one-page addendum, Subbaswamy peeled off the outreach dimension and the emphasis on democracy studies, zeroed in on the new undergraduate major in American Studies, and kept the budget where it was. "What got funded," the dean wrote to the incoming director (Guterl), "bears no resemblance to what was proposed." In the whirlwind of shuffling through a dozen or so such proposals, the college merely returned a massive windfall to American Studies, charged the director to create a new degree, and left it to us to define the intellectual features of the major. The new base budget became available in the fall of 2005.

This wasn't just good luck. Until 2005, the American Studies program enjoyed a fine reputation as a small, niche graduate program on a very big campus, a sort of "switching point" for the humanities, for the College of Arts and Sciences, and for the campus. Historically, it emerged in the late 1950s during the internationalization of the university, largely as a programmatic unit, bringing guest speakers to campus, organizing symposia and conferences, and encouraging co-teaching. Our earliest records indicate that the program was shaped profoundly by the arrival of folklorist Richard Dorson in 1957. Consistent with national trends, the establishment of an American Studies curriculum was originally intended to provide graduate students with unique, layered, and genuinely interdisciplinary or multidisciplinary viewpoints on American culture, politics, and society. The first courses were team-taught by several faculty members, many from English, history, and folklore. Approved in 1964, the program's combined PhD—ambitious and flexible—allowed students pursuing a PhD in the college (or a terminal doctoral-level degree in other units anywhere on campus, including in law, education, journalism, and music) to concentrate their additional coursework in a complementary degree, with a new focus in their writing and research. In short, the program had a solid reputation, a long record of faculty commitment, and good success with a complicated degree that was measured in decades.

Although we did not have an undergraduate major, in 2003 we had recently added a fifteen-credit minor to our curriculum; we also offered a wide range of undergraduate courses that both served our minors and allowed other students in the college to fulfill their distribution requirements.[5] The new undergraduate major was thus poised to add a stronger, deeper foundation to this venerable edifice. The centerpiece of the program's transformation was a unique, hybrid curriculum that engaged with issues of interdisciplinarity and transnationalism that are at the heart of current debates in American Studies, Latin American Studies, area studies, and elsewhere.[6] The central premise of the major—and now the program in general—is that the boundaries of American Studies are "the Americas" writ large, and that the hemisphere is an illuminating context for understanding the history, culture, and social formations of the United States. The curriculum did not overlook either the importance of national history—and of smaller, regional, and local histories within that—or the profound influence of Africa, Asia, and Europe in the making of the United States. The faculty's expertise in these areas comes through in multiple ways. American Studies regularly offers courses such as "The Image of America in the World," "American Studies in Transnational Contexts," and "Foreign Studies in American Studies" alongside "Comparative American Identities" and "Indigenous Worldviews in the Americas." In recent years, these courses have been taught by Americanist faculty from African American and African Diaspora Studies, English, History, Spanish and Portuguese, and other such units whose research centers on the United States, Latin America, and the Caribbean, and the relations between these regions and also with Africa, Asia, Oceania, and so on. Thus, our curriculum offered students a window into both the hemispheric and the transnational dimensions of the newest, most expansive visions of American Studies.

The use of the hemispheric as an analytic is on prominent display in all of the core classes, including the entry-level "gateway" course—once too cleverly titled, "Democracy in the Americas," but now retitled as a simple, if troublesome, question: "What Is America?" One standard feature of the class is an early unit that brings together Jack Kerouac's *On the Road*, fragments of the English translation of Ernesto "Che" Guevara's so-called "Motorcycle Diaries," the Walter Salles film (2005) of the same name, and Ridley Scott's *Thelma and Louise* (1991).[7] The point of this unit is to interrogate the assumptions of utopian revolutionary politics, and to put that careful interrogation of those compelling assumptions at the heart of the class. It is also to stress, from the very beginning, that this class and, by extension, this major are not bordered in the usual ways. That is, we want the students in "What Is America?" to understand that the radicalism of "Che," "Sal," "Dean," "Thelma," and "Louise" is produced by a certain hemispherics, where all radical roads lead to something beyond the nation-state. Students write a long essay on the imaginative role of "the Americas" in each of these texts, but they also labor to connect them, to highlight common themes, patterns, and experiences. Subsequent readings, discussions, and essay assignments build on this unit, and focus the class on the overlapping, dissonant meanings of "America." In this course, as in the other core courses, we want them to recognize right away that "American Studies" will push them outside of the United States, and that our conception of "American" is hemispheric and comparative.

What made this formulation of American Studies possible, institutionally speaking, was a legacy of the Cold War, the source of the institution's greatest expansion. Removed from any major cities and thousands of miles from any ocean, the university had scrambled during the 1950s and 1960s to establish itself as a national resource for foreign language, study abroad, and global cultural education. This effort was enormously successful, and these areas continue to be a strength of the university. Indeed, IU offers more than eighty "foreign" languages, and has more Title VI centers than any other public university in the country. But the programs and centers that once trained (and still do, occasionally) the nation's "striped pants men" and covert operatives have also bored holes sideways through the disciplines. As a campus, we have more *interdisciplinary* units on campus than we do *disciplinary* units, and many of them, courtesy of the U.S. Department of Education, have budgets in the millions of dollars. We have *hundreds* of faculty attached to these units, each a laboratory for cross-disciplinary or interdisciplinary work. This meant, in a practical sense, that when we sought to bring American Studies scholars to campus and put them in closer conversation with academics from Latin America, we had the support of the Center for Latin American and Caribbean Studies, and the financial flexibility that comes from Title VI funds. It also meant that our focus on interdisciplinarity was neither strange nor new. Indeed, so much of the critical initial support for our expansion into undergraduate education was drawn from area studies and foreign language units that one department chair in the traditional humanities joked that we seemed intent on building up Spanish enrollments.

The degree was intended to be cutting-edge, with an emphasis on the hemispheric, the transnational, and the global, and our rapid start-up was meant to facilitate a leapfrogging over "older" and more traditional models in place elsewhere. Many American Studies BA degrees, we deduced from a careful survey of our peers, had multiple tracks at the undergraduate level, and lots of "either/or" formulations in the early years of a student's coursework. There are good reasons for this flexibility, but we thought to make the most of it by insisting on a single common sequence of courses, and then by encouraging individualized foci that would lead up to—and provide intellectual support for—the senior seminar experience. The major integrates intensive seminar-style courses, interdisciplinary research methods, and an emphasis on critical reading skills to provide a well-rounded, interdisciplinary, liberal arts education. Fifteen credits are devoted to a rigid sequence of required courses that provides a common curricular experience for all majors. The sequence includes "*What Is America?,*" *our wide-open gateway course that engages with the histories and cultures of the Americas as a whole;* "*Comparative American Identities*"; "*Topics in Interdisciplinary American Studies*"; "*American Studies in Transnational Contexts*"; *and a capstone senior seminar. Together, this sequence constitutes one of* the first genuinely interdisciplinary concentrations offered nationwide in the field.

On the one hand, it bridges the social sciences and humanities, whereas most other degree programs offer students a choice of "tracks" that mirror the divide between these areas. On the other hand, it comprises a narrative "through line" that provides intellectual cohesion and assists in the process of forging community among our students.

With their remaining fifteen credits, students work with the director to design a concentration of elective and cross-listed courses that allows them to pursue their specific academic interests within American Studies. This dialogue with the director is a serious feature of a student's entry into the degree, and nearly every student chooses a concentration that is global, comparative, or hemispheric. Still, these range from the universal ("Racism and Oppression" or "Border Zones") to the obscure ("Music and American Politics"). For nearly every student involved, the concentration is meant to be background coursework for the senior thesis written in the final course in the core sequence, the senior seminar in American Studies.

We installed other features to complement the basic degree, highlighting the "postnational" feel of American Studies at IU. Many of our courses integrate an intensive writing component. And as a way of underscoring the imbrication of the American nations with each other and with the world at large, as well as speaking to the diversity within each one, we require students to complete three years of language study—one year more than the college's standard language requirement. In addition, students are allowed to count courses on related subject matter taught in languages other than English towards their requirements for the major. Finally, we have an enhanced emphasis on study abroad. The intensive writing courses, the language requirement, and the inclusion of courses in other languages in our curriculum help to ensure academic rigor, a global character, and intellectual seriousness, while "What Is America?" and the flexible concentration appeal to larger numbers of students, drawing them into the major.

Implementing the major required that we have faculty who would be regularly available to staff courses and mentor students, which, in turn, meant that we could no longer rely solely on our affiliated faculty in other units. Making the major successful further required what Phil Deloria, following Michael Goldberg, calls "articulated interdisciplinarity," wherein individual faculty members must stretch, or reach, across unit lines.[8] In 2005, we began the first of several searches that eventually brought eight new junior faculty members—a total of four FTE—into our ranks. The emphasis in each search was placed on joint appointments with other units in the college. This was, in part, a structural requirement: As a program, we cannot serve as a tenure home, and thus we needed to seek out partners who could provide this shelter to our hires. However, this need actually worked in tandem with our aspirations of interdisciplinary hires, ensuring that our hires would bridge different units and maintain their commitment to both disciplinary and interdisciplinary practice, even as it allowed us to maximize the breadth of specialties that could be brought to IU and ensure that basic college hiring needs were met. The emphasis on topical breadth was critically important, given the program's desire to hire a balanced community of scholars with interests in both the United States and Latin America.

The aim of these joint appointments was to build bridges between the various units that were close partners with American Studies. In all preliminary conversations with the college and potential partner departments, we stressed the importance of such "articulated interdisciplinarity." Thus, we were able to work together to identify and recruit top-notch interdisciplinary scholars whose research encompassed more than one narrowly defined specialty, and who would thrive in a multiple appointment setting. It was no accident that we made five of our seven hires with interdisciplinary units (i.e., African American and African Diaspora Studies, Communication and Culture, and Religious Studies). We were keenly aware that joint appointments place additional pressures on faculty, especially when they were pre-tenure, as was the case with all of our hires. In order to protect them, then, from the very beginning, we also worked with our hiring partners to establish both short- and long-term expectations for how research, teaching, and service were to be divided between the units. We were able to present these expectations to the candidates from the interview stage, and upon a faculty member's arrival on campus, we codified them with what is known at IU as a "Memorandum of Understanding."

In the long run, we have been extremely fortunate: The tensions that American Studies as a discipline has experienced elsewhere with other fields as it has moved towards transnational scholarship were virtually absent from our own experiences, another unanticipated courtesy of the Cold War shaping of the Bloomington campus. Our emphasis on the new world hemispherics of "America" has not been seen as aggressive by the college's sizeable cohort of Latin Americanists. Indeed, when the Title VI grant held by our Center for Latin American and Caribbean Studies (CLACS) was revised in 2007, we were asked to submit agenda items for inclusion in the budget, a request that has helped to keep alive our connection to the University of Guyana in Georgetown. And in 2008, we co-sponsored with CLACS a successful conference on "Blackness in Latin America and the Caribbean." Finally, one of our original faculty appointments in American Studies was hired with Title VI funds from the U.S. Department of Education, channeled through CLACS, while another position was created for our program for a Latin Americanist after we lobbied the dean in conjunction with CLACS and our International Studies program. The creation of an American Studies degree hasn't been seen as a threat to English or to History: Because we hire faculty jointly and have shared standards for tenure, we are viewed by many as a perfect complement. Likewise, because students at IU often double-major or even triple-major, we aren't seen as a threat to major growth for other units. And with so many jointly appointed faculty on campus, our personnel profile is hardly distinctive. In short, the academic architecture of the Bloomington campus—with some of the nation's largest and oldest Title VI centers creating wide bore holes sideways through the college, and foregrounding multidisciplinary and interdisciplinary scholarship across standard units—has made it easier for us to position American Studies as "just another" area studies.

Graduate Ends

The implementation of the undergraduate major has had a ripple effect through our graduate program. American Studies offers a twelve-credit PhD minor, but the majority of our graduate students are enrolled in our combined PhD program, in which students pursuing a doctorate in another department take additional coursework and qualifying exams in American Studies, and write a dissertation with committee members representing both units. There is a structural imbalance here, for we do not have a say in the admissions process: Students must first be admitted to a PhD program in the home department, after which we admit them into our program. Students also receive the bulk of their funding from their home departments.

The synergy between our undergraduate and graduate programs made it important to revisit the graduate curriculum and incorporate a significant transnational component in the program of study. Graduate students teach in our undergraduate program, and whether they are developing their own courses or teaching "What Is America?," they must be able to position their subject matter in relation to the undergraduate major's hemispheric frame. Beyond this pedagogical rationale, however, it was important to us to make our requirements more rigorous, implementing changes that would train graduate students both to think about American Studies hemispherically and, more generally, to locate themselves within the discipline of American Studies as easily and as well as they do in their home departments. The students in the combined PhD program have traditionally sought jobs in their home disciplines rather than in American Studies. Our students are enthusiastic about and committed to American Studies, but have tended, not surprisingly, to prioritize the requirements and pressures of the home department through which they were admitted to the university and from which they receive virtually all of their funding. We wanted to change this to make American Studies a vital part of how they approach their scholarship and view themselves as scholars, so we set out to revise the curriculum to better prepare them to present their research at the American Studies Association convention and to be strong candidates for positions in American Studies.

Our first step was to increase the number of core courses that students take in the program. As our students are already enrolled in PhD programs, American Studies has historically been flexible in its requirements in order to facilitate their progress and time to degree. And the prior emphasis has overwhelmingly been on interdisciplinary and multidisciplinary work. When we assumed our current positions, students in the combined PhD program were only required to take two core courses, totaling eight credits out of the thirty-two that we require: "Introduction to American Studies" (G603) and a variable-topic research seminar (G751). Their remaining credits came from electives from the courses that our affiliate faculty teaches in their respective home departments and list with us, as well as from American Studies-related topics in the students' home department. This meant that students in the program had very little common ground, either methodologically or thematically. To provide more consistent training, we revised the content of another course, "Perspectives in American Studies" (G604), traditionally taught as a methodologies course, to focus on transnational and hemispheric issues; in the fall of 2006, the faculty voted to make G604 a third required course and to increase from sixteen to twenty the number of credits that students must take in American Studies courses outside of their home departments. In effect, the expansion of the faculty for the undergraduate program that was enabled by the CTE funds benefitted the graduate program by allowing us to implement this requirement: Whereas in the past our course offerings were essentially limited to what our affiliated faculty were offering in their home departments, having faculty with FTE (and, therefore, regular teaching commitments) in our program gave us more control over curricular planning and allowed us to schedule core courses more regularly.

In addition to increasing course requirements, we implemented a more rigorous qualifying exam as part of the American Studies degree requirement (students, of course, also take exams in their home departments). The exam, which is written, consists of two parts: one question on the institutional contexts and intellectual history of American Studies, and one that probes how students' discipline-based research both benefits from and contributes to the work of American Studies. The exam structure in effect when we started had students prepare, in consultation with the associate director and their exam committee, a reading list of approximately thirty works for the exam, drawing on a list of recommended works that had been devised by American Studies faculty. The exam structure was last revised in 2003, with the suggested reading list implemented as a response to faculty interest in avoiding the creation of a fixed and rigid canon. We felt, however, that its merits in preparing students for their research projects by allowing them to tailor their readings in American Studies to their research interests were overshadowed by its inability to provide students with a common and in-depth understanding of the field as a whole. To remedy this, our graduate studies committee proposed that students prepare two lists: a mandatory one consisting of more than 120 works, and one consisting of approximately 50 works that bridges their disciplinary interests and American Studies. This approach provides both depth in American Studies preparation and conceptual tools that allow students to integrate their work in their home department with their American Studies interests. This exam structure went into effect in January 2008, and we are currently seeing the first group of students move towards taking the new exam. It is, of course, a far more demanding exam than what was previously in place, and it may lead some students to choose the PhD minor rather than the combined PhD. These issues were discussed when faculty met to vote on the new reading list: In much the same way as we sought to implement an "honors-style" undergraduate major, we also wanted to attract students to the combined PhD who were committed to the degree and who were willing to take on the additional workload, as these were the students who would best be able to work towards job placement in either their home departments or American Studies units. The new reading list was approved unanimously by faculty, and we have, indeed, been able to attract students to the combined PhD program at a rate similar to what it was under the old exam.

The master list provides students with an overview of the discipline, its emergence and history, as well as representative methodologies and groundbreaking scholarship. Having a common list also serves broader functions within our program. First, it provides students with an opportunity to work together as they prepare for the exam, helping them to build a sense of community that the graduate program has long lacked. Second, we have asked faculty who teach the core courses in the program to draw some of their readings from it: This both allows us to frame our curriculum and eases the burden of the heavier reading load on students. Third, it clarifies—in a very good way—the difference between the philosophical expectations of their home departments and those of American Studies. In some ways, though, the list is still a work in progress in ways that attest to some of the difficulties faced by the discipline of American Studies as it has gone transnational: We have, for example, yet to fully find a way of taking into consideration work in languages other than English, and the list would benefit from the inclusion of more work by scholars outside of the United States, as well as by Latin Americanists and others doing inter-American research. Given that so few doctoral programs here in Bloomington require serious and intensive language study—despite the presence of world-class resources—this last issue might be insurmountable. Unless, that is, American Studies were to institute its own language requirement. We continue, for now, to dwell on the structural limitations before us, though we do plan to include more research by Latin Americanists when we next revise the list.

In addition to these curricular changes, the graduate program has been profoundly affected by the aforementioned expansion of our faculty. This expansion has been literal, for we have hired eight core faculty members and a lecturer during this period. But it has also been figurative in many ways as well. In keeping with the hemispheric emphasis of the program, we have increased our geographic range: Four of our recent hires are Latin Americanists, and we have partnered with both IU's Center for Latin American and Caribbean Studies, a Title VI National Research Center, and our rapidly growing International Studies program to bring in several of these hires; and when Cohn became associate director in 2005, she was the first faculty member from the Department of Spanish and Portuguese in our ranks, and was soon followed by several more. For much of its existence, American Studies at IU (utterly dependant on faculty volunteerism) was dominated by Folklore, History, and English, but the new hires and additions from within have allowed us to further our transnational mission, for they have formed the basis of good working relationships with other internationally oriented academic units.

Though sharing the Cold War origins of the area studies, American Studies has not generally been considered an area studies program. Area studies and languages are traditional strengths of IU, which has more Title VI research centers than any other public university in the United States. But while there was a long history of collaboration among the area studies programs, the Cold War compartmentalization of these interdisciplinary units resulted in few interactions between these programs and American Studies—a situation that has changed over the past few years here. The presence of internationally oriented faculty has attracted other faculty interested in doing transnational research to our program. We have, for example, added several Latin Americanists as affiliate faculty, as well as others whose work falls within other area studies programs. This has allowed us to collaborate on research projects with these units. We have also established partnerships for teaching courses with International studies, for example, and regularly cross- and joint-list courses with Latin American and Caribbean Studies, African American and African Diaspora Studies, Latino Studies, and Spanish and Portuguese.

The active role that this faculty has taken in the program has expanded our geographic scope, as well as our ability to navigate the tensions inherent in bringing area studies disciplines together to do transnational and collaborative research and programming. We do not intend to single out these faculty members' contributions while overlooking those of our other affiliated faculty: We have been extremely fortunate to have a dedicated faculty who, across the board, have devoted considerable time and effort to the program, and we would not have been able to accomplish any of these changes without their commitment. We simply mean to acknowledge within the context of this discussion that it is absolutely critical to our goals to have faculty—whether new arrivals or long-standing affiliates—who are trained in transnational methodologies, both in order to share their expertise with students through teaching and mentoring and to face head on the infrastructural challenges brought about by our growing numbers of students developing dissertation topics with a transnational dimension.

In the past year alone, students from the Department of Spanish and Portuguese have, for the first time, enrolled in our graduate programs, and several more students from other departments are doing research in border studies within our program. Comparative Literature, too, has generated as many combined PhD students in recent years as History or English. As this cohort of students grows, it will fundamentally alter the way the American Studies program and faculty work: We may well face dissertations written in other languages, and we certainly will face research projects whose scope lies beyond the boundaries of the United States that may well test the limits of our competencies. We are learning on the front lines, then, just how much the seismic shift in American Studies brought about by the transnational turn affects not just the subject and methods of analysis, but also poses a fundamental challenge to received models of intellectual training, evaluation, research and publication, and curricular development, transforming the very infrastructure within which scholarship is produced.

We have also sought to couple these new theoretical models with a move towards training graduate students beyond the bounds of the U.S. academy, largely through the development of faculty-led, intensive, summer study-abroad courses—"Afro-Guyana," "Dominican Identity and Race in the Age of Globalization," and "Black-Irish Connections" (scheduled for summer 2011)—for graduates and undergraduates. We encourage (and provide financial support for) our graduate students and faculty to attend the "usual" American Studies workshops held in the summer at Dartmouth and at the Clinton Institute, but we are also an institutional sponsor and routine participant in the Tepoztlán Institute in Mexico. On a thematic level, these initiatives link the study of race, race making, nation, and globalization. On a professional level, they are critical to the students' training, for they introduce students to scholars, artists, and political figures from beyond the United States as peers and colleagues engaged in parallel scholarly inquiry. This is an important way of helping the students to establish connections and, in the process, providing them with tools that will allow them to move their research outside the bounds of the U.S. academy, to engage with scholarship produced in venues that traditionally have lain outside the bounds of American Studies, and to facilitate their access to work done outside of the United States, as well as that done in languages other than English. It is also an effort to get beyond mere abstract clichés, and to make transnational exchange a practical feature of research and writing.

The emphasis on the hemispheric and the global opened other strange doors, some of them rather unexpectedly. Occasionally, we were included as one of the area studies in campus discussions. Several times, we were asked to provide lecturers for State Department "U.S. Studies Institutes" held in Africa, and several American Studies faculty members participated via teleconferencing or in person. Occasionally, we are asked for something beyond what we know or understand—for example, being given two weeks to write a $500,000 grant proposal for a historically black fraternity looking to make inroads in South Africa. And more than once the program has played host to visiting delegations from Algeria, Tunisia, and China, or has given a home to visiting scholars funded either by Fulbright or by the State Department. Indeed, in our small office we have done everything from setting up bank accounts to arranging for housing and signing up children for school, often without the benefit of a common tongue. IU's federally supported emphasis on international connection and language acquisition brings foreign scholars here in large numbers, and when they come to our program, they share space and staff with us; attend our workshops, conferences, and job talks; and shape our conversations around the table. For the same reasons we strive to get our students into contact with their colleagues outside of the United States, we just don't say "no" to these requests.

Another significant—even surprising—consequence of the new undergraduate major and the amplification of the graduate program has been the development of a PhD minor in Native American and Indigenous Studies (NAIS), which went into effect in the fall of 2008. The hiring plan for the undergraduate major sparked an intense faculty discussion about hiring an expert in native history or literature, two vacant specialties here in recent years. And from that debate came a proposal for a doctoral minor, bringing together a distinctive community of faculty, many of them in the social sciences. The geographic scope of the minor is deliberately broad, encompassing Native American communities throughout the Americas as well as indigenous populations throughout the world. An explicit goal of the minor, in fact, is to "place the study of American Indians within the context of a broader, more sweeping and international inquiry into the nature of political power, colonial settlement, and global contact" (http://www.indiana.edu/~amst/graduates/native.shtml). This emphasis shadows the overall focus of the American Studies program on hemispheric, global, and comparative issues, a parallelism that has been institutionally and academically helpful. It has allowed us to bring in faculty who are South Africanists and Latin Americanists, and to welcome students who study the First Peoples of Canada, the Ainu of Japan, and the Aborigines of Australia. It has focused our attention on translation, on cultural circulation, and on interchange and connection around the world. And it links us to recent scholarship on indigeneity, which has drawn from postcolonial studies to broaden the frame of inquiry for those who once studied natives peoples in tribal or national contexts.

Of course, the relationship between American Studies and Native American Studies has historically been fraught with tension.[9] Within Native American Studies, there is a particular concern about the hegemonic interests of the larger field and a regular disdain for the label of "ethnic studies." There is also a jarring, as yet underexamined, intellectual dissonance between the critical emphasis in Native American Studies on sovereignty and the flow and direction of what has been called "postnational" American Studies. Indeed, from some angles, it can seem as if American Studies is losing interest in the nation-state at precisely the moment when indigenous peoples have good cause to be studying it closer. Our explicit emphasis—in American Studies and in NAIS—on the global, the hemispheric, and the comparative thus highlighted a common perspective, a shared view of the relationship between the particular and the transnational.

Such partnerships, of course, have long been the object of administrative obsession, offering a chance to save a few dollars while folding together a small clutch of interdisciplinary units. This is especially true in the current budget crisis, when any astute administrator can refer to leading American Studies programs at NYU, USC, and Michigan, each of which features various "ethnic" concentrations or foci. Here at IU, budget issues were compounded by a Responsibility-Centered Management (RCM) system that ultimately laid bare a practical gap between our efforts to implement an "honors-style" degree and the need to demonstrate our immediate "accountability" through the hasty recruitment of majors. If we intended to create something serious and slow to build, the marketplace mentality of our RCM campus was invoked in the very first year we offered the BA. We were also quickly steered towards what was labeled "amalgamation" with several smaller Ethnic Studies units, politically affiliated programs that had only just earned the right to exist on their own terms. In short, just as soon as IU's American Studies program offered its own BA degree, its role as a switching point for the campus was no longer just a curricular strength but now also a political weakness.

Indeed, it is astonishing how quickly the relationship between these parallel initiatives—the growing presence of American Studies, the emergent NAIS community, and the new ethnic studies programs on campus—was transformed into a debate about consolidation. When the NAIS faculty, acting with the support of the American Studies program, created a basic governance agreement and elected their own director (who would serve as an associate director of American Studies), a late spring budget surprise stopped the process. Over subsequent months, the amicable relationship between American Studies and NAIS became a mathematical proof that "the studies units" could be collapsed into one big jumble. Any positive expression of that cross-disciplinary affiliation resurrected old conversations about what was variously named "amalgamation," "confederation," and "consolidation." A joint proposal for a shared graduate degree facilitated by many different interdisciplinary programs and departments quickly became an opportunity for the administration to formally propose a merger of these units, offering new digs that would involve a drastic reduction in space for all, as well as a displacement from the center of the campus. Forcefully, politely, and at some cost, American Studies and NAIS demurred.

Conclusions

Interdisciplinarity, we learned the hard way, is not merely an abstraction, or an intellectual practice reflected in writing. It isn't just about engaging other disciplines or fields in an exchange, or dialogue in which both sides accommodate or compromise. It is also about teaching students, modeling new and challenging modes of thought for them, and turning them loose on the world. If we limit our attention just to writing, or public performance, or toggling back and forth between this and that approach in our conference papers and monographs, we focus on the *easiest* manifestations of what we earlier called a new/nouvelle/nuevo American Studies. Indeed, as difficult as it is to work through a single research question that crosses continents, oceans, and languages, it is even harder to change budget-driven institutions that have served for so long as repositories of nationalist knowledge and as engines of knowledge production for the nation-state, and to do so in ways that challenge American exceptionalism. But these same new approaches to American Studies bring us much closer than ever before to area studies programs, to the languages and literatures of the Americas, and to international studies, religious studies, comparative literature, Jewish studies, and other object-focused, globally oriented fields. It may be that outside of the wealthiest and most determined private institutions, this thing we call "transnational" or "hemispheric" American Studies is more likely to flourish at the mammoth institutions built up by the American Century, and to use the architecture left behind by the Cold War to make this new intellectual practice manifest.

To conclude, then, we'd like to offer three lessons for anyone seriously interested in offering a transnationally or hemispherically oriented American Studies degree.

First, take the study of "foreign" languages very, very seriously. Despite its commitment to transnationalism, the American Studies Association is still fundamentally monolingual—at its conferences, in its publications, etc. We recognize the challenges of accommodating other languages within the organization's structure, but this type of subtle change is both compatible with the "biggest tent" mentality of American Studies, and it counters some of the skepticism about the ASA's transnational intentions that are felt by scholars working in and with other languages. This isn't an easy charge. Our goal is to train our students—especially at the undergraduate level—to feel that language study is as much a part of their degree as are courses in, say, U.S. history, so that working in multiple languages becomes an integral part of both their methodology and their view of the Americas. But we've struggled to implement this, especially at the graduate level, where our program status as "the perfect complement" has made it difficult to break away from the basic requirements of "home" departments. And we are mindful that working closely and thoroughly in a language other than English can never be the *sine qua non* of American Studies.

Second, our new partners move beyond the usual suspects to include the area studies centers and programs. There will be quirks: the Department of Education, for example, considers "Latin America" to end at Mexico's northern border, so Title VI funds may not be used to fund collaborations looking at, say, immigration or Latino communities. Title VI and area studies centers also tend to have a greater emphasis on quantitative work and greater representation from the social sciences, neither of which corresponds to current American Studies practice. But these are minor obstacles. We have worked together at the level of hiring, courses, conferences, speakers, and so on, and these collaborations have greatly expanded our audience and, ultimately, our reach. In its new transnational, global, or hemispheric incarnations, American Studies might well profit from much closer solidarities with programs, departments, centers, and institutes that have utterly irrefutable historical connections to American foreign policy agendas in the period between 1947 and 1989, but that now might have newer—or at least broader—agendas and missions. And we have our own Cold War origins story, too, despite our desire to establish a history that is too happily traced back to radical social movements.[10]

Third, we think it is important to know your institution. The truth is that we went into this as idealists, but left as realists. If we'd set out to hire eight "transnationalists," all of whom had published in *American Quarterly*, we would have run into a series of political roadblocks and the major wouldn't have worked. We had a goal—build a new American Studies major—but institutional factors forced us to think (and rethink) locally about allies and opportunities. Our goal was theorized through our own research, our experience at the ASA, and our readings of colleagues and peers, but we brought that theorization to the ground in Bloomington by thinking carefully about the history of our program and the nature of our college and our campus. Our American Studies program, in short, makes perfect sense *here*. It makes sense elsewhere, too, but our attention to local politics—and to the kinds of crawlspaces that were available when we went looking for more breathing room—was extremely important.

Notes

1. The classic expression is, of course, Janice A. Radway, "What's in a Name? Presidential Address to the American Studies Association, 20 November 1998," *American Quarterly* 51.1 (March 1999): 1–32.
2. For more on hemispherics, see Caroline Levander and Robert Levine, eds., Hemispheric American Studies (New Brunswick, NJ: Rutgers University Press, 2007).
3. John Muthyala, Reworlding America: Myth, History, and Narrative (Athens: Ohio University Press, 2006).
4. At this time and, indeed, until we made our first hire in 2006, the program had no faculty with permanent FTE.
5. The College of Arts and Sciences at IU requires that students take a certain number of credits in the arts and humanities, the social and historical sciences, and the natural and mathematical sciences.
6. For a sampling of the literature that informed our thinking, see Michelle Ann Stephens, Black Empire: The Masculine Global Imaginary of Caribbean Intellectuals in the United States, 1914–1962 (Durham: Duke University Press, 2005); Micol Seigel, Uneven Encounters: Making Race and Nation in Brazil and the United States (Durham: Duke University Press, 2008); Martha Hodes, The Sea Captain's Wife: A True Story of Love, Race, and War in the Nineteenth Century (New York: W. W. Norton, 2006); Kirsten Silva Gruesz, Ambassadors of Culture: The Transamerican Origins of Latino Writing (Princeton: Princeton University Press, 2002); Brent Hayes Edwards, The Practice of Diaspora: Literature, Translation, and the Rise of Black Internationalism (Cambridge: Harvard University Press, 2003); Ifeoma Kiddoe Nwankwo, Black Cosmopolitanism: Racial Consciousness and Transnational Identity in the Nineteenth Century (Philadelphia: University of Pennsylvania Press, 2005); Kevin K. Gaines, African American in Ghana: Black Expatriates in the Civil Rights Era (Chapel Hill: University of North Carolina Press, 2006); Mary Dudziak, Cold War Civil Rights: Race and the Image of American Democracy (Princeton: Princeton University Press, 2000); Brenda Gayle Plummer, ed., Window on Freedom: Race, Civil Rights, and Foreign Affairs, 1945–1988 (Chapel Hill: University of North Carolina, 2003); Penny Von Eschen, Race Against

Empire: Black Americans and Anti-Colonialism (Ithaca: Cornell University Press, 1997) and Satchmo Blows Up the World: Jazz Ambassadors Play the Cold War (Cambridge: Harvard University Press, 2005); Adam McKeown, Chinese Migration Networks and Cultural Change: Peru, Chicago, Hawaii: 1900-1936 (Chicago: University of Chicago Press, 2001); Matthew Frye Jacobson, Special Sorrows: The Diasporic Imagination of Irish, Polish, and Jewish Immigrants in the United States (Cambridge: Harvard University Press, 2001) and, more generally, Thomas Bender, ed., Rethinking American History in a Global Age (Berkeley: University of California Press, 2002).

7. The original idea for this unit sprang from a reading of María Josefina Saldaña-Portillo, "On the Road with Che and Jack: Melancholia and Colonial Geographies of Race in the Americas," New Formations 47 (Summer 2004): 87–108.

8. Philip J. Deloria, "Broadway and Main: Crossroads, Ghost Roads, and Paths to an American Studies Future," American Quarterly 61.9 (March 2009): 4.

9. Robert Allen Warrior, "A Room of One's Own at the ASA: An Indigenous Provocation," American Quarterly 55.4 (December 2003): 681–687; Philip Deloria, "American Indians, American Studies, and the ASA," American Quarterly 55.4 (December 2003): 669–680; Elizabeth Cook-Lynn and Craig Howe, "The Dialectics of Ethnicity in America: A View from American Indian Studies," in Johnella Butler, ed., Color-Line to Borderland: The Matrix of American Ethnic Studies (Seattle: University of Washington Press, 2001), 150–168.

10. See, for instance, George Lipsitz's, American Studies in a Moment of Danger (Minneapolis: University of Minnesota Press, 2001).

Works Cited

Abernethy, Francis E. "Dobie, James Frank (1888–1964)." *The Handbook of Texas Online.* http://www.tshaonline.org/handbook/online/articles/DD/fdo2.html (accessed 21 May 2009).
———. *The Texas Folklore Society.* Vol.1. Denton: University of North Texas Press, 1992.
Abu-Lughod, Janet L. Before European Hegemony: *The World System a.d. 1250–1350.* New York: Oxford University Press, 1989.
Acuña-Soto, Rodolfo, David W. Stahle, Malcolm K. Cleaveland, and Matthew D. Therrell. "Megadrought and Megadeath in 16th Century Mexico." *Emerging Infectious Diseases* 8.4 (April 2002): 360–375.
Adams, James Truslow. *The Epic of America.* Boston: Little, Brown, and Company, 1931.
Adams, Rachel. *Continental Divides: Remapping the Cultures of North America.* Chicago: University of Chicago Press, 2010.
Adorno, Rolena. *Writing Resistance in Colonial Peru.* Austin: University of Texas Press, 1986.
Agee, James and Walker Evans. *Let Us Now Praise Famous Men.* New York: Mariner Books, 2001.
Aiton, Arthur Scott. *Antonio de Mendoza: First Viceroy of New Spain.* Durham: Duke University Press, 1927.
Alba, Richard and Victor Nee. *Remaking the American Mainstream: Assimilation and Contemporary Immigration.* Cambridge, MA: Harvard University Press, 2003.
Alegría, Fernando. *Walt Whitman en Hispanoamérica.* Mexico City: Fondo de Cultura Económica, 1954.
Alonzo, Amanda C. *Tejano Legacy: Rancheros and Settlers in South Texas, 1734–1900.* Albuquerque: University of New Mexico Press, 1998.
Anzaldúa, Gloria. *Borderlands/La Frontera: The New Mestiza.* San Francisco: Aunt Lute Books, 1987.
Arrighi, Giovanni. *The Long Twentieth Century.* London: Verso, 1994.
Bailey, Julius H. *Around the Family Altar: Domesticity in the African Methodist Episcopal Church, 1865–1900.* Gainesville: University Press of Florida, 2005.
Bannon, John Francis. *Bolton and the Spanish Borderlands.* Norman: University of Oklahoma Press, 1964.

Bartlett, Thomas. "Archive Fever." *The Chronicle of Higher Education.* July 20, 2007.

Baudot, Georges. Introduction to *Historia de los Indios de la Nueva España,* by Fray Toribio de Motolinia. Madrid: Clásicos Castalia, 1985.

Baur, K. Jack. "Mexican War." *The Handbook of Texas Online.* http://www.tshaonline.org/handbook/online/articles/MM/qdm2.html (accessed 6 May 2009).

Bauer, Ralph. *The Cultural Geography of Colonial American Literatures: Empire, Travel, Modernity.* New York: Cambridge University Press, 2003.

———. "Hemispheric Studies." *PMLA* 124.1 (January 2009): 234–245.

———. "Early American Literature and American Literary History at the 'Hemispheric Turn.'" *American Literary History* (Summer 2010): 250–266.

Bender, Steven W. "Direct Democracy and Distrust: The Relationship Between Language Law Rhetoric and the Language Vigilantism Experience." *Harvard Latino Law Review* 2 (2005): 145–174.

Bender, Thomas, ed. *Rethinking American History in a Global Age.* Berkeley: University of California Press, 2002.

Bernstein, Richard J. *Radical Evil: A Philosophical Investigation.* Cambridge, UK: Polity, 2002.

Bloom, Lisa. "Introducing With Other Eyes." In *With Other Eyes: Looking at Race and Gender in Visual Culture,* edited by Lisa Bloom, 1–16. Minneapolis: University of Minnesota Press, 1999.

Blumer, Herbert. "The Future of the Color Line." In *The South in Continuity and Change,* eds., J. C. McKinney and E. W. Thompson. 322–336. Durham, NC: Duke University Press, 1965.

Bolívar, Simón. "Carta de Jamaica." In *Fuentedes de la Cultura Ladtinoamericana, compiled by Leopold Zae.* Vol 1. Mexico: Fonda de Cultura Económica, 1993 [1815].

Bolton, Herbert E. "The Epic of Greater America," *American Historical Review* 38 (April 1933): 448–474.

———. *History of the Americas: A Syllabus with Maps.* Boston: Ginn and Company, 1928.

———. *The Spanish Borderlands: A Chronicle of Old Florida and the Southwest.* New Haven: Yale University Press, 1921.

———. *Wider Horizons of American History.* 1939. Notre Dame: University of Notre Dame Press, 1967.

Bonilla-Silva, Eduardo. "From Biracial to Triracial: Towards a New System of Racial Stratification in the USA." *Ethnic and Racial Studies* 27.6 (2004): 931–950.

———. *Racism without Racists: Color-Blind Racism and the Persistence of Racial Inequality in America.* 3rd ed. Lanham, MD: Rowman & Littlefield, 2009.

Bonilla-Silva, Eduardo and Karen Glover. "We Are All Americans: The Latin Americanization of Race Relations in the USA." In *The Changing Terrain of Race and Ethnicity: Theory, Methods and Public Policy,* eds., A. E. Lewis and M. Krysan. 149–183. New York: Russell Sage, 2004.

Borah, Woodrow Wilson. Introduction to *Secret Judgments of God: Old World Disease in Colonial Spanish America*. Norman: University of Oklahoma Press, 1991.

Bordo, Susan. *Unbearable Weight: Feminism, Western Culture, and the Body*. Berkeley: University of California Press, 1993.

Bornholdt, Laura. "The Abbi de Pradt and Monroe Doctrine." The *Hispanic American Historical Review* xxiv (1944): 201–21.

Brading, David A., *The First America: The Spanish Monarchy, Creole Patriots and the Liberal State, 1492–1867*. Cambridge: Cambridge University Press, 1993.

Brickhouse, Anna. *Transamerican Literary Relations and the Nineteenth-Century Public Sphere*. New York: Cambridge University Press, 2004.

Brooks, Francis J. "Revising the Conquest of Mexico: Smallpox, Sources, and Populations." *Journal of Interdisciplinary History* 24.1 (Summer, 1993): 1–29.

———. Christopher R. N. DeCorse, and John Walton, eds. *Small Worlds: Method, Meaning, and Narrative in Microhistory*. Santa Fe: School for Advanced Research Press, 2008.

Buckley, Liam. "Studio Photography and the Aesthetics of Citizenship in the Gambia, West Africa." In *Sensible Objects: Colonialism, Museums, and Sensible Objects,* edited by Edwards, Gosden, Phillips, 61–85. Oxford: Oxford University Press, 2006.

Bulmer, M. G. *Francis Galton: Pioneer of Heredity and Biometry*. Baltimore: John Hopkins University Press, 2003.

Bustamante, Jorge. "Trafficking and Vulnerability of Migrants: A Conceptual Form." *Mexicans Look at Mexico* Lecture Series. 20 November 2007.

Callahan, Monique-Adelle. *Between the Lines*. New York: Oxford University Press, forthcoming, 2011.

Campos, Roy. "Todos Blanos…Menos Los Que No." *Nexos* 31 (2009): 64–65.

Cañizares-Esguerra, Jorge. *Puritan Conquistadors: Iberianizing the Atlantic, 1550–1700*. Stanford: Stanford University Press, 2006.

Carbado, Devon W. "Racial Naturalization." *American Quarterly* 57.3 (2005): 633–658.

Carby, Hazel V. *Race Men*. Cambridge, MA: Harvard University Press, 1998.

Chevigny, Bell Gale, and Gari Laguardia, eds. *Reinventing the Americas: Comparative Studies of Literature of the United States and Spanish America*. Cambridge: Cambridge University Press, 1986.

Clarke, Kamari Maxine and Deborah A. Thomas, eds. *Globalization and Race*. Durham: Duke University Press, 2006.

Clarke, Grahm. *The Photograph: A Visual and Cultural History*. New York: Oxford UP, 1997.

Coakley, Sarah. "Introduction." In *Religion and the Body*, edited by Coakley, 2–10. New York: Cambridge University Press, 1997.

Coates, Ta-Nehisi. "Is Obama Black Enough?" *Times*. Feb. 2007.http://www.time.com/time/nation/article/0,8599,1584736,00.html (accessed 11 January 2010).

Cole, John Y. *Jefferson's Legacy: A Brief History of the Library of Congress*. Washington: Library of Congress, 1993.

Connexions. Rice University. <http://cnx.org> and <http://cnx.org/lenses/oaap>.

Cook, Noble David and W. Gorge Lovell, *Secret Judgments of God: Old World Disease in Colonial Spanish America*. Norman: University of Oklahoma Press, 1991.

Cook, Sherburne F. and Woodrow Borah. *Essays in Population History: Mexico and the Caribbean*. Vol. 1. Berkeley: University of California Press, 1971.

Cook-Lynn, Elizabeth and Craig Howe. "The Dialectics of Ethnicity in America: A View from American Indian Studies." In *Color-Line to Borderland: the Matrix of American Ethnic Studies,* edited by Johnella Butler, 150–168. Seattle: University of Washington Press, 2001.

Cotera, María Euginia. "A Woman of the Borderlands: 'Social Life in Cameron, Starr, and Zapata Counties' and the Origins of Borderlands Discourse." In *Life Along the Border: A Landmark Tejana Thesis*. College Station, TX: Texas A&M University Press, 2006.

———. "Jovita González Mireles: A Sense of History and Homeland." In *Latina Legacies: Identity, Biography, and Community*, edited by Vicki Ruiz and Virginia Korrol. New York: Oxford University Press, 2005.

———. Native Speakers: Ella Deloria, Zora Neale Hurston, Jovita González, and the Poetics of Culture. Austin: University of Texas Press, 2008.

Cottrell, Debbie Mauldin. "Casis, Lilia Mary (1869–1947)." In *The Handbook of Texas Online*. http://www.tshaonline.org/handbook/online/articles/CC/fcace_print.html (accessed 6 May 2009).

Cox, Christoph. "On Evil: An Interview of Alenka Zupancic." *Cabinet* 5 (Winter 2001–2002).

Crosby, Alfred W. *The Columbian Exchange: Biological and Cultural Consequences of 1492*.Westport: Greenwood Press, 1972.

Cunningham, Michael. *Crowns: Portraits of Black Women in Church Hats*. New York: Doubleday, 2000.

Daniels, Roger. *Guarding the Golden Door: American Immigration Policy and Immigrants Since 1882*. New York: Hill and Wang, 2004.

Dash, J. Michael. *The Other America: Caribbean Literature in a New World Context*. Charlottesville: University of Virginia Press, 1998.

Dayan, Joan. *Haiti, History and the Gods*. Berkeley: University of California Press, 1998.

De las Casas, Bartolomé. *A Short Account of the Destruction of the Indies,* edited and translated by Nigel Griffin, with an introduction by Anthony Pagden. London: Penguin Classics, 1992.

———. *Brevísima relación de la destruyción de las Indias*. Introduction by Miguel León Portilla. Madrid: EDAF, 2004.

DeParle, Jason and Robert Gebeloff. "Food Stamp Use Soars, and Stigma Fades." *New York Times*. 28 Nov. 2009. http://www.nytimes.com/2009/11/29/us/29foodstamps.html?emc=eta1 (accessed 15 January 2010).

Deloria, Philip J. "American Indians, American Studies, and the ASA." *American Quarterly* 55.4 (December 2003): 669–680.

———. "Broadway and Main: Crossroads, Ghost Roads, and Paths to an American Studies Future." *American Quarterly* 61.9 (March 2009): 1–25.

Deloria, Vine. *God Is Red: A Native View of Religion*. Colorado: Fulcrum Publishing, 1993 [1972].
Derrida, Jacques. *Archive Fever: A Freudian Impression*, translated by Eric Prenowitz. Chicago: University of Chicago Press, 1995.
———. *Mal d'Archive:Une Impression Freudienne*. Paris: Galilée, 1995.
Dimock, Wai Chee and Lawrence Buell, Eds. *Shades of the Planet: American Literatureas World Literature*. Princeton: Princeton University Press 2007.
———. *Through Other Continents: American Literature Across Deep Time*. Princeton: Princeton University Press, 2006.
Dobie, J. Frank. To Mrs. Jovita González de Mireles. 27 April 1937. The University of Texas at Austin.
Douglas, Karen Manges. "A Voice from the Grave: A Posthumous Dialogue with Norma Williams." *Studies in Symbolic Interaction*. 32 (2008): 197–216.
Doyle, Don and Pamplona, Marco Antonio Pamplona, eds. *Nationalism in the New World*. Athens: University of Georgia Press, 2006.
Du Bois, W. E. B. *The Souls of Black Folk*. New York: Vintage Books, 1990 [1904].
Dudziak, Mary. *Cold War Civil Rights: Race and the Image of American Democracy*. Princeton: Princeton University Press, 2000.
Durand, Jorge, and Douglas S. Massey. "What We Learned from the Mexican Migration Project." In *Crossing the Border: Research from the Mexican Migration Project*, edited by Jorge Durand and Douglas S. Massey. 1–14. New York: Russell Sage Foundation, 2004.
Dussel, Enrique. *The Invention of the Americas: Eclipse of "the Other" and the Myth of Modernity*, trans. Michael D. Barber. New York: Continuum, 1995.
Dyreson, Mark. "The 'Physical Value' of Races and Nations: Anthropology and Athletics at the Louisiana Purchase Exposition." In *The 1904 Anthropology Days and Olympic Games: Sport, Race, and American Imperialism*, edited by Susan Brownell. Critical Studies in the History of Anthropology Series. Lincoln, NE: University of Nebraska Press, 2008.
Early American Digital Archive. Maryland Institute for Technology in Humanities at the University of Maryland. http://www.mith2.umd.edu/eada.
Edwards, Brent Hayes. *The Practice of Diaspora: Literature, Translation, and the Rise of Black Internationalism*. Cambridge, MA: Harvard University Press, 2003.
Elliott, J. H. *Empires of the Atlantic World, 1492–1830*. New Haven: Yale University Press, 2006.
EssenceHat.com. http://www.essencehat.com/ (accessed July 5, 2009).
Fanning, Patricia J. *Through an Uncommon Lens: The Life and Photography of F. Holland Day*. Amherst: University of Massachusetts Press, 2008.
Feagin, Joe R. *Systemic Racism: A Theory of Oppression*. New York: Routledge, 2006.
Fernández-Armesto, Felipe. *The Americas: A Hemispheric History*. New York: Modern Library, 2003.

Firmat, Gustavo Pérez. *Do the Americas Have a Common Literature?* Durham: Duke University Press, 1990.

Fitz, Earl E. *Inter-American Literature and Criticism: An Electronic Annotated Bibliography.* <http://www.uiowa.edu/uiowapress/interamerican/>.

———. "Old World Roots/New World Realities: A Comparatist Looks at the Growth of Literature in North and South America." *Council on National Literatures/Quarterly World Report* 3.3 (July 1980): 8–11.

———. *Rediscovering the New World: Inter-American Literature in a Comparative Context.* Iowa City: University of Iowa Press, 1991.

Florescano, Enrique. *Memory, Myth, and Time in Mexico.* Austin: Austin University Press, 1994.

Fong, Eric and Rima Wilkes. "Racial and Ethnic Residential Patterns in Canada." *Sociological Forum.* 18. 4 (2003): 577–602.

Fosdick, Raymond. *The Story of the Rockefeller Foundation.* New York: Harper & Row, 1952.

Foucault, Michel. *The Archeology of Knowledge.* New York: Pantheon, 1972.

———. *Discipline and Punish: The Birth of the Prison.* New York: Vintage Books, 1979.

———. *The History of Sexuality.* Vol. 1–3. New York: Vintage Books, 1978–1986.

Foucault, Michel. *Les mots et les choses.* París: Gallimard, 1967. Spanish edition: *Las palabras y las cosas.* México: Siglo XXI Editores, 1968.

Fox, Claire F. "Comparative Literary Studies of the Americas." *American Literature* 76.4 (2004): 871–86.

———, ed. *Critical Perspectives and Emerging Models of Inter-American Studies.* Spec. issue of Comparative American Studies 3.4 (2005): 387–532.

Gabara, Esther. *Errant Modernism: The Ethos of Photography in Mexico and Brazil* Durham: Duke University Press, 2008.

Gaines, Kevin K. *African Americans in Ghana: Black Expatriates and the Civil Rights Era.* Chapel Hill: University of North Carolina Press, 2006.

Gallop, Jane. "The Historicization of Literary Studies and the Fate of Close Reading." *Profession 2007.* 181–186. New York: MLA, 2007.

Garza-Falcón, Leticia Magda. *Gente Decente: A Borderlands Response to the Rhetoric of Dominance.* Austin: University of Texas Press, 1998.

Gaskins, Pearl Fuyo. *What Are You? Voices of Mixed-Race Young People.* New York: Henry Holt & Co, 1999.

General Terms and Conditions for Discretionary Grants from the Institute of Museum and Library Services. <http://wwww.imls.gove/recipients/administration.shtm> and <http://www.neh.gov/manage/gtcao.html#foreign>.

Gibson, Charles. *The Aztecs under Spanish Rule: A History of the Indians of the Valley of Mexico, 1519–1810.* Stanford: Stanford University Press, 1964.

Giles, Paul. "Commentary: Hemispheric Partiality," in "Hemispheric American Literary History." *American Literary History,* edited by Caroline Levander and Robert Levine (Fall 2006) 18.3: 648–656.

Gillman, Susan. "Otra Vez Caliban/Encore Caliban: Adaptation, Translation, Americas Studies." *American Literary History* (Spring/Summer 2008): 187–209.

Gilroy, Paul. *The Black Atlantic: Modernity and Double Consciousness.* Cambridge, MA: Harvard University Press, 1993.

Glenn, Evelyn. "Citizenship and Inequality: Historical and Global Perspectives." *Social Problems* 47.1 (2000):1–20.

Glissant, Edouard. *Poetics of Relation*, translated by Betsy Wing. Ann Arbor: University of Michigan Press, 1997 [1990].

Gomez, Michael, ed. *Diasporic Africa: A Reader.* New York: New York University Press, 2006.

González, John. "Terms of Engagement: Nation or Patriarchy in Jovita González's and Eve Raleigh's *Caballero.*" In *Recoverying the U.S. Hispanic Literary Heritage*, edited by José F. Aranda, Jr. and Silvio Torres-Saillant. Vol. 4, 264–276. Houston: Arte Público Press, 1993.

González, Jovita. "America Invades the Border Towns." *Southwest Historical Review* 15 (1930): 469–477.

———. "Among My People." *Publications of the Texas Folklore Society* 8 (1930): 99–108.

———. Jovita González de Mireles Papers. Special Collections & Archives. Texas A&M University-Corpus Christi. Mary and Jeff Bell Library. Corpus Christi, TX.

———. "The Folklore of the Texas-Mexican Vanquero." *Publications of the Texas Folklore Society* 4 (1927): 7–22.

———. *Life Along the Border: A Landmark Tejana Thesis*, María Euginia Cotera, edited. Elma Dill Russell Spencer Series in the West and Southwest. College Station, TX: Texas A&M University Press, 2006.

González de Mireles, Jovita and Margaret Eimer (Eve Raleigh). 30 May 1939. Texas State University, San Marcos.

———. "Tales and Songs of the Texas-Mexicans." *Publications of the Texas Folklore Society* 8 (1930): 109–114.

———. *The Woman Who Lost Her Soul and Other Stories*, edited by Sergio Reyna. Recovering the U.S. Hispanic Literary Heritage series. Houston: Arte Público Press, 2000.

———. *Untitled Handwritten Autobiography.* Texas A&M University-Corpus Christi.

Grider, Sylvia Ann. "Dorothy Scarborough." In *Texas Women Writers: A Tradition of Their Own*, edited by Sylvia Ann Grider and Lou Halsell Rodenberger. College Station, TX: Texas A&M University Press, 1997.

Gruesz, Kirsten Silva. *Ambassadors of Culture: The Transamerican Origins of Latino Writing.* Princeton: Princeton University Press, 2002.

Gruesz, Kirsten Silva. "Translation: A Key(word) into the Language of America(nists)." *American Literary History* 16 (2004): 85–92.

Gruzinski, Serge. *La colonisation de l'imaginaire: Sociétés indigenes et occidentalisation dans le Mexique espagnol XVI–XVIII siècle.* Paris: Gallimard, 1988. (Spanish edition: *La colonización de lo imaginario: Sociedades indígenas y occidentalisation en el Mexico español Siglos XVI–XVII.* Mexico: Fonda de Cultura Económica, 1995.)

Gunder, Frank A. and Barry K. Gills, eds. *The World System: Five Hundred Years or Five Thousand?* London: Routledge, 1993.

Guimond, James. *American Photography and the American Dream.* Chapel Hill: University of North Carolina, 1991.

Gutierrez, David. "Significant to Whom? Mexican Americans and the History of the American West." *Western Historical Quarterly* 24.4 (Nov. 1993): 519–539.

Gutiérrez, Ramón A. and Genaro M. Padilla. Introduction to *Recovering the U.S. Hispanic Literary Heritage.* Houston: Arte Público Press, 1993.

Guterl, Matthew. *American Mediterranean: Southern Slaveholders in the Age of Emancipation.* Boston: Harvard University Press, 2008.

Haines, Michael and Richard Steckel, eds. *The Population History of North America.* Cambridge: Cambridge University Press, 2000.

Hamilton, Kenda. "The Dialect Dilemma: Whether One Is Speaking Ebonics or Appalachian English, Sociolinguists Say All Dialects Are Created Equal." *Black Issues in Higher Education* 22.5 (2005): 34–36.

Handley, George. *Postslavery Literature in the Americas: Family Portraits in Black and White.* Charlottesville: University of Virginia Press, 2000.

———. *New World Poetics: Nature and the Adamic Imagination in Whitman, Neruda, and Walcott.* Athens: University of Georgia Press, 2007.

Hanke, Lewis. "The Development of Latin American Studies in the United States, 1939–1945." *The Americas* 4 (1947): 32–64.

———, ed. *Do the Americas Have a Common History? A Critique of the Bolton Theory.* New York: Knopf, 1964.

"Harlem's Heaven Hat Boutique." <http://www.harlemsheaven.com/church-hats> (accessed July 5, 2009).

Harvey, Bruce. *American Geographics: U.S. National Narratives and the Representation of the Non-European World, 1830–1865.* Stanford: Stanford University Press, 2001.

Heller, Dick D. Jr., "Roma-Los Saenz, Texas." In *The Handbook of Texas Online*, http://www.tshaonline.org/handbook/online/articles/RR/hgr6_print.html (accessed 6 May 2009).

Henry, Marsha Giselle. "'Where Are You Really From?': Representation, Identity and Power in Fieldwork Experiences of a South Asian Diasporic." *Qualitative Research* 3 (2003): 229–242.

Hight, Eleanor M. and Gary D. Sampson. "Introduction: Photography, 'Race,' and Post-Colonial Theory." In *Colonialist Photography: Imag(in)ing Race and Place*, edited by Hight and Sampson, 1–19. New York: Routledge, 2002.

Hodes, Martha. *The Sea Captain's Wife: A True Story of Love, Race, and War in the Nineteenth Century.* New York: W. W. Norton, 2006.

Hochschild, Jennifer L. and Brenna Marea Powell. "Racial Reorganization and the United States Census 1850–1930: Mulattoes, Half-Breeds, Mixed Parentage, Hindoos, and the Mexican Race." *Studies in American Political Development* 22 (2008): 59–96.

Huntington, Samuel. *The Clash of Civilizations and the Remaking of World Order.* New York: Simon & Schuster, 1999.

"Immigration to Mexico." http://en.wikipedia.org/wiki/Immigration_to_Mexico (accessed November 10, 2009).
Instituto Mora. 2010. <http://www.institutomora.edu.mx/>.
International Council on Archives. http://www.ica.org/.
Jacobson, Matthew Frye. *Special Sorrows: The Diasporic Imagination of Irish, Polish, and Jewish Immigrants in the United States.* Cambridge, MA: Harvard University Press, 2001.
James, Angela. "Making Sense of Race and Racial Classification." In *White Logic, White Methods: Racism and Methodology,* eds., T. Zuberi and E. Bonilla-Silva. 31–36. Lanham, MD: Rowman & Littlefield, 2008.
Jasso, Guillermina. "Whom Shall We Welcome? Elite Judgments of the Criteria for the Selection of Immigrants." *American Sociological Review* 53.6 (1988): 919–932.
Jefferson, Thomas [1984]. Cited in Michel-Rolph Trouillot. *Silencing the Past: Power and the Production of History.* Boston: Beacon Press, 1995.
———. *The Writings of Thomas Jefferson,* edited by A. A. Lipscomb. Vol. 13. Washington, D.C.: Library of Congress, 1903–1904 [1813].
Jensen, Robert. *The Heart of Whiteness: Confronting Race, Racism, and White Privilege.* San Francisco: City Lights, 2005.
Jiménez, Blas R. *El Nativo: Versos en Cuenos Para Espantar Zombies.* Santo Domingo: Editora, Búho, 1996.
———. *Caribe Africano en Despertar.* Santo Domingo: Editora Manatí, 2006.
———. *Afrodomicano por Eleccion Negro Por Nacimiento.* Santo Domingo: Editora Manatí, 2008.
Johnson, Benjamin Heber. *Bordertown: The Odyssey of an American Place.* New Haven: Yale University Press, 2008.
Johnson, Michael P. "Denmark Vesey and His Co-Coonspirators." *The William and Mary Quarterly* 58.4 (2001): 915–976.
Kadir, Djelal. Introduction. *America: The Idea, the Literature.* Ed. Djelal Kadir. Spec. issue of *PMLA* 118.1 (2003): 9–24.
Kamiya, Gary. "Cablinasian Like Me." *Salon* (April 1997). http://www.salon.com/april97/tiger970430.html (accessed 11 January 2010).
Katznelson, Ira. *When Affirmative Action Was White: An Untold History of Racial Inequality in Twentieth-Century America.* New York: W. W. Norton, 2005.
Kaup, Monika. "The Unsustainable *Hacienda*: the Rhetoric of Progress in Jovita González and Eve Raleigh's *Caballero*." *Modern Fiction Studies* 51:3 (2005): 561–591.
Keen, Maurice. "Main Currents in the United States Writings on Colonial Spanish America, 1884–1984." Hispanic American Historical Review 65 (1985): 657–682.
Kissinger, Henry. *Years of Renewal.* New York: Simon & Schuster, 1999.
Klor de Alva, Jorge. 1992. "The Postcolonization of (Latin) American Experience: A Reconsideration of 'Colonialism,' 'Postcolonialism,' and 'Mestizaje.'" Reprinted in: Gyan Prakash. *After Colonialism: Imperial Histories and Postcolonial Displacements.* Princeton: Princeton University Press, 1995.
Kreneck, Thomas H. Preface to *Crossing the Rio Grande: An Immigrant's Life in the 1880s.* ix–xii. College Station: Texas A&M University Press, 2007.

Kuhn, Annette and Kirsten Emiko McAllister. "Locating Memory: Photographic Acts—An Introduction." In *Locating Memory: Photographic Acts, edited by Kuhn and McAllister*, 1–17 New York: Berghahn Books, 2006.

Kutzinski, Vera M. *Against the American Grain: Myth and History in Williams Carlos Williams, Jay Wright, and Nicolás Guillén*. Baltimore: Johns Hopkins University Press, 1987.

Langley, Lester. *The Americas in the Age of Revolution: 1750–1850*. New Haven: Yale University Press, 1998.

Landau, Paul S. "Empires of the Visual: Photography and Colonial Administration in Africa." In *Images and Empires: Visuality in Colonial and Postcolonial Africa*, edited by Landau and Deborah D. Kaspin, 141–171. Berkeley: University of California Press, 2002.

Langford, Martha. "Speaking the Album: An Application of the Oral-Photographic Framework." In *Locating Memory: Photographic Acts*, edited by Kuhn and McAllister, 223–246. New York: Berghahn Books, 2006.

Lanning, John Tate. *Academic Culture in the Spanish Colonies*. London: Oxford University Press, 1940.

Latapí, Augustín Escobar, and Susan F. Martin, eds. *Mexico-U.S. Migration Management: A Binational Approach*. Lanham, MD: Lexington Books, 2008.

Lazo, Rodrigo. *Writing to Cuba: Filibustering and Cuban Exiles in the United States*. Chapel Hill: University of North Carolina Press, 2005.

Lein, Laura and Deanna T. Schexnayder, Karen Manges Douglas and Daniel Schroeder. *Life After Welfare: Reform and the Persistence of Poverty*. Austin, TX: University of Texas Press, 2007.

León, Arnoldo De. *They Called Them Greasers: Anglo Attitudes Toward Mexicans in Texas, 1821–1900*. Austin: University of Texas Press, 1983.

Leonard, Irving A. *Baroque Times in Old Mexico: Seventeenth-Century Persons, Places, and Practices*. Ann Arbor: University of Michigan Press, 1959.

Levander, Caroline F. and Robert S. Levine, eds. *Hemispheric American Literary History*. Spec. issue of *American Literary History* 18.3 (2006): 397–655.

———. *Hemispheric American Studies*. New Brunswick: Rutgers University Press, 2008.

Levine, Robert. *Dislocating Race and Nation: Episodes in Nineteenth-Century American Literary Nationalism*. Chapel Hill: University of North Carolina Press, 2008.

Lewis, Bernard. *The Shaping of the Modern Middle East*. New York: Oxford University Press, 1997.

Limón, José, ed. *Dew on the Thorn*. Houston: Arte Público Press, 1997.

———. *Dancing with the Devil: Society and Cultural Poetics in Mexican-American South Texas*. Madison, WI: University of Wisconsin Press, 1994.

Lipsitz, George. *American Studies in a Moment of Danger*. Minneapolis: University of Minnesota Press, 2001.

"List of Countries by Population." http://en.wikipedia.org/wiki/Country_population (accessed November 4, 2009).

Livi-Bacci, Massimo. "The Depopulation of Hispanic America after the Conquest." *Population and Development Review* 32.2 (June 2006): 199–232.

Lockhart, James. *The Nahuas after the Conquest*. Stanford: Stanford University Press, 2000.

Lockhart, James and Stewart B.Schwartz. *Early Latin America, a history of colonial Spanish America and Brazil*. Cambridge Latin American Studies. Cambridge: Cambridge University Press, 1983.

Lockwood, Cara. *Dixieland Sushi*. New York: Downtown Press, 2005.

López, Ian F. Haney. *White by Law: The Legal Construction of Race*. New York: New York University Press, 1996.

MacCormack, Sabine. *Religion in the Andes: Vision and Imagination in Early Colonial Peru*. Princeton: Princeton University Press, 1991.

Magnaghi, Russell. *Herbert E. Bolton and the Historiography of the Americas*. Westport: Greenwood Press, 1998.

Major, John. *Prize Possession: The United States Government and the Panama Canal 1903–1979*. Cambridge: Cambridge University Press, 2003.

Mancillas, Aida, Ruth Wallen, and Marguerite R. Waller. "Marking Art, Making Citizens: Las Comadres and Postnational Aesthetics." In *With Other Eyes: Looking at Race and Gender in Visual Culture,* edited by Lisa Bloom, 107–132. Minneapolis: University of Minnesota Press, 1999.

Manoff, Marlene. "Theories of the Archive from Across the Disciplines." *Portal: Libraries and the Academy* 4 (2004), 9–25.

Manríquez, B. J. "Argument in Narrative: Tropology in Jovita González's *Caballero*." *Bilingual Review/La revista bilingue* 25:2 (2000): 172–178.

Márquez, Lourdes and Andrés del Angel. "Height Among the Prehispanic Mayas of the Yucatan peninsula." *Bones of the Mayas, Studies of Ancient Skeletons,* edited by Whittington, Stephen and David Reed. Washington, D.C.: Smithsonian Institute Press, 1997.

Marr, John S. and James B. Kiracofe. "Was the Huey *Cocoliztli* a Haemorragic Fever? *Medical History* 44 (2000): 341–362.

Martin, Joyce A., Brady E. Hamilton, Paul D. Sutton, Stephanie J. Ventura, Fay Menacker, Sharon Kirmeyer, and T. J. Mathews. *Births: Final Data for 2006*. Hyattsville, MD: National Center for Health Statistics, 2009.

Massey, Douglas S. *Categorically Unequal: The American Stratification System*. New York: Russell Sage Foundation, 2008.

———. "Why Does Immigration Occur: A Theoretical Synthesis." In *The Handbook of International Migration*, edited by Charles Hirschman, Philip Kasinitz, and Josh DeWind. 34–52. New York: Russell Sage Foundation, 1999.

Mauad, Ana Maria. "Composite Past: Photography and Family Memories in Brazil (1850–1950)." In *Art and the Performance of Memory: Sounds and Gestures of Recollection,* edited by Richard Cándida Smith, 215–233. New York: Routledge, 2002.

McCaa, Robert. "Spanish and Nahuatl Views on Smallpox and Demographic Catastrophe in Mexico." *Journal of Interdisciplinary History* 25.3 (Winter 1995): 397–431.

———. "The Peopling of Mexico from Origins to Revolution." In *The Population History of North America*. 241–304. New York: Cambridge University Press, 2000.

McClennen, Sophia. "Inter-American Studies or Imperial American Studies?" *Comparative American Studies* 3.4 (2005): 393–413.

McQuire, Scott. *Vision of Modernity: Representation, Memory, Time and Space in the Age of the Camera*. Thousand Oaks, CA: Sage Publications, 1998.

Melville, Elinore G. K. *A Plague of Sheep: Environmental Consequences of the Conquest of Mexico*. Cambridge: Cambridge University Press, 1994.

Mendoza, Louis Gerard. *Historia: The Literary Making of Chicana and Chicano History*. College Station: Texas A&M University Press, 2001.

Mignolo, Walter. *The Darker Side of the Renaissance: Literacy, Territoriality, & Colonization*. Ann Arbor: University of Michigan Press, 2003.

Morieras, Alberto. *The Exhaustion of Difference: The Politics of Latin American Cultural Studies*. Durham: Duke University Press, 2001.

Moretti, Franco. "Conjectures on World Literature." *New Left Review* 1 (Jan.–Feb. 2000): 54–68.

Morgan, David. *Visual Piety: A History and Theory of Popular Religious Images*. Berkeley: University of California Press, 1999.

Morgan, Jennifer L. "Male Travelers, Female Bodies, and the Gendering of Racial Ideology, 1500–1700." In *Bodies in Contact: Rethinking Colonial Encounters in World History*, edited by Tony Ballantyne and Antoinette Burton, 54–66. Durham: Duke University Press, 2005.

Montejano, David. *Anglos and Mexicans in the Making of Texas, 1836–1986*. Austin: University of Texas Press, 1987.

Motolinía, Fray Toribio de. *History of the Indians of New Spain,* translated and annotated by Francis Borgia Steck, with a Bio-bibliographical Study. Washington, D.C.: Academy of American Franciscan History, 1951.

———. *Historia de los Indios de la Nueva España*. Introduction by Georges Baudot. Madrid: Clásicos Castalia, 1985.

Murga, Aurelia Lorena. "Morena/o." In *International Encyclopedia of the Social Sciences,* ed. W. A. Darity. 2nd ed. Vol. 4. 293–295. Detroit: Macmillan Reference USA, 2008.

Muthyala. John. *Reworlding America: Myth, History, and Narrative*. Athens: Ohio University Press, 2006.

National Endowment for the Humanities: Grant Management. General Terms and Conditions for Awards to Organizations. May 2009. <http://www.neh.gov/manage/gtcao.html#foreign>.

National Study on the Use of Libraries, Museums and the Internet. 2006. <http://interconnectionsreport.org >.

Ngai, Mae M. "The Architecture of Race in American Immigration Law: A Reexamination of the Immigration Act of 1924." *Journal of American History*. 86 (1999): 167–192.

———. *Impossible Subjects: Illegal Aliens and the Making of Modern America*. Princeton, NJ: Princeton University Press, 2004.

Nwankwo, Ifeoma Kiddoe. *Black Cosmopolitanism: Racial Consciousness and Transnational Identity in the Nineteenth Century*. Philadelphia: University of Pennsylvania Press, 2005.

O'Gorman, Edmundo. "Do the Americas Have a Common History?" Hanke 103–111.

———. *The Invention of America: An Inquiry into the Historical Nature of the New World and the Meaning of Its History*. Bloomington: Indiana University Press, 1961.

O'Hare, Peggy. "Face of Food Stamp Program Changing in Texas." *Houston Chronicle*. 11 Jan. 2010. http://apps.grouptivity.com/socialmail/main.do?uId=234274&tId=455635&pk=120086374337&acn=zj!d9&pId=HeOHCWXaPRs=&acn=zj!d9 (accessed on 11 January 2010).

Orfield, Gary and Susan E. Eaton. *Dismantling Desegregation: The Quiet Reversal of Brown V. Board of Education*. New York: W. W. Norton, 1997.

Orvell, Miles. *American Photography*. New York: Oxford University Press, 2003.

Osborne, Thomas. "The Ordinariness of the Archive." *History of the Human Sciences* 12 (May 1999), 51–55.

Our Americas Archive Partnership: Hemispheric Approach to Inter-American Archives. Institute of Museum and Library Science. Rice University. <http://oaap.rice.edu>.

Paredes, Américo. *Folklore and Culture on the Texas-Mexican Border*, edited by Richard Bauman. Austin: University of Texas Press, 1993.

———. *A Texas-Mexican Cancionero: Folksongs of the Lower Border*. Urbana: University of Illinois Press, 1976.

Parker, Hershel. *Herman Melville: A Biography*. Vol. 1. Baltimore: Johns Hopkins University Press, 1996.

Passel, Jeffrey S. "Unauthorized Migrants: Numbers and Characteristics." Washington, D.C.: Pew Hispanic Center, 2005.

Patterson, Tiffany Ruby and Robin D. G. Kelley. "Unfinished Migrations: Reflections on the African Diaspora and the Making of the Modern World." *African Studies Review* 43.1 (Special Issue on the Diaspora, April 2000): 11–45.

Pérez Firmat, Gustavo, ed. *Do the Americas Have a Common Literature?* Durham: Duke University Press, 1990.

Pinn, Anthony. *Embodiment and the New Shape of Black Theological Thought*. New York: New York University Press, 2010.

Pinn, Anthony. "Introduction" and "A Beautiful *Be-ing*: Religious Humanism and the Aesthetics of a New Salvation." *Black Religion and Aesthetics: Religious Thought and Life in Africa and the African Diaspora*, edited by Anthony Pinn. New York: Palgrave Macmillan, 2009.

Pinn, Anthony. *Terror and Triumph: The Nature of Black Religion*. Minneapolis: Fortress Press, 2003.

Picca, Leslie Houts and Joe R. Feagin. *Two-Faced Racism: Whites in the Backstage and Frontstage*. New York: Routledge, 2007.

Plummer, Brenda Gayle, ed. *Window on Freedom: Race, Civil Rights, and Foreign Affairs, 1945–1988*. Chapel Hill: University of North Carolina, 2003.

Pool, William C. "Barker, Eugene Campbell (1874–1956)." *The Handbook of Texas Online*, http://www.tshaonline.org/handbook/online/articles/BB/fba65.html (accessed 21 May 2009).

———. *Eugene C. Barker, Historian*. Austin: Texas State Historical Association, 1971.

Portales, Marco. *Crowding Out Latinos: Mexican Americans in the Public Consciousness*. Philadelphia: Temple University Press, 2000.

Porterfield, Nolan. *Last Cavalier: The Life and Times of John A. Lomax, 1867–1948*. Chicago: University of Illinois Press, 2001.

Prem, Hanns J. "Disease Outbreak during the Sixteenth Century." In *'Secret Judgments of God': Old World Disease in Colonial Spanish America*. Noble David Cook and W. Gorge Lovell. Norman: University of Oklahoma Press, 1991.

Prescott, William H. *History of the Conquest of Mexico*. New York: Modern Library, 2001.

———. *History of the Conquest of Peru*. New York: Modern Library, 1998.

ProEnglish Action. "English in the 50 States." Arlington, VA. ProEnglish Action. 2010. http://www.proenglish.org/issues/offeng/states.html (accessed 25 January 2010).

Quadagno, Jill. *The Color of Welfare: How Racism Undermined the War on Poverty*. New York: Oxford University Press, 1996.

Quick, Andrew. "The Space Between: Photography and the Time of Forgetting in the Work of Willie Doherty." In *Locating Memory: Photographic Acts*, edited by Kuhn and McAllister, 162–171. New York: Berghahn Books, 2006.

Radway, Janice A. "What's in a Name? Presidential Address to the American Studies Association, 20 November 1998."*American Quarterly* 51.1 (March 1999): 1–32.

Raleigh, Eve and González Jovita. *Caballero: A Historical Novel*. College Station, TX: Texas A&M University Press, 1996.

Ramirez, Pablo. "A Borderlands Response to American Eugenics in Jovita González and Eve Raleigh's *Caballero*." *Canadian Review of American Studies*. 39.1 (2009): 21–39.

Robles, Frances. "Black Denial." *Miami Herald*. 13 June 2007. http://www.miamiherald.com/multimedia/news/afrolatin/part2/index.html (accessed 13 January 2010).

Roberts, Julian V. and Anthony N. Doob. "Race, Ethnicity, and Criminal Justice in Canada." *Crime and Justice* 21:469–522, 1997.

Rodríguez, J. Javier. "Caballero's Global Continuum: Time and Place in South Texas." *MELUS: The Journal of the Society for the Study of the Multi-Ethnic Literature of the United States* 33.1 (2008 Spring): 117–138.

Rojo, Antonio Benítez. James Maraniss, Translator. *The Repeating Island: The Caribbean and the Postmodern Perspective.* Durham: Duke University Press, 1997.

Romolini, Jennifer. "Louisiana Judge Denies Interracial Couple Marriage License—Why Are We Surprised?" *Shine* (19 Oct. 2009). http://shine.yahoo.com/channel/sex/louisiana-judge-denies-interracial-couple-marriage-license-why-are-we-surprised-527014/ (accessed 11 January 2010).

Ruggles, Steven, Matthew Sobek, Trent Alexander, Catherine A. Fitch, Ronald Goeken, Patricia Kelly Hall, Miriam King, and Chad Ronnander. *Integrated Public Use Microdata Series: Version 4.0* [Machine-readable database]. Minneapolis, MN: Minnesota Population Center, 2010 [producer and distributor]. http://usa.ipums.org/usa/cite.shtml (accessed on 10 January 2010).

Russell, Rucks. "Some Blacks Say 'Negro' on U.S. Census is Offensive, Outdated during Modern Times." *khou.com* 8 Jan. 2010. http://www.khou.com/home/khou-81038557.html (accessed 11 January 2010).

Saenz, Rogelio. "Latinos and the Changing Face of America." In *The American People: Census 2000*, eds. R. Farley and J. Haaga. 352–379. New York: Russell Sage Foundation, 2005.

Saenz, Rogelio and Maria Cristina Morales. "Demography of Race and Ethnicity." In *The Handbook of Population*, eds. D. L. Poston Jr. and M. Micklin. 169–208. New York: Klewer Academic/Plenum Publishers, 2005.

Saenz, Rogelio, Karen Manges Douglas, David Geronimo Embrick, and Gideon Sjoberg. "Pathways to Downward Mobility: The Impact of Schools, Welfare, and Prisons on People of Color." In *Handbook of the Study of Racial and Ethnic Relations*, eds. H. Vera and J. R. Feagin. 373–409. New York: Springer, 2007.

Saldaña-Portillo, María Josefina. "On the Road with Che and Jack: Melancholia and Colonial Geographies of Race in the Americas." *New Formations* 47 (Summer 2004): 87–108.

Saldívar, José David. *The Dialectics of Our America: Genealogy, Cultural Critique, and Literary History.* Durham: Duke University Press, 1991.

Saldívar, Ramón. *The Borderlands of Culture: Américo Paredes and the Transnational Imaginary.* Durham: Duke University Press, 2006.

San Miguel, Guadalupe Jr. *"Let All of Them Take Heed": Mexican Americans and the Campaign for Educational Equality in Texas, 1910–198.* Austin: University of Texas Press, 1988.

Sánchez, José R. "Dead Latinos." *National Institute for Latino Policy Commentary.* 2010. http://www.authorslatino.com/blog/archives/121 (accessed on 15 January 2010).

Santamarina, Xiomara. "'Are We There Yet?'": Archives, History, and Specificity in African-American Literary Studies." *American Literary History* 20 (Spring/Summer 2008): 304.

Scott, Rebecca. *Degrees of Freedom: Louisiana and Cuba after Slavery.* Cambridge, MA: Harvard University Press, 2008.

Seigel, Micol. *Uneven Encounters: Making Race and Nation in Brazil and the United States.* Durham: Duke University Press, 2008.

Sekula, Allan. "The Body and the Archive." In *The Body: A Reader*, edited by Mariam Fraser and Monica Greco, 163–166. New York: Routledge, 2005.

Sen, Rinku. "As American as Apple Pie (and Samosas and Tacos)." *New America Media*. 4 July 2008. http://news.newamericamedia.org/news/view_article.html?article_id=862f0c45f28b95df20cda13b9d76aea5 (accessed 11 January 2010).

Shilling, Chris. *The Body and Social Theory*. 2nd ed. London: Sage Publications, 2003.

Shell, Marc and Werner Sollors. *The Multilingual Anthology of American Literature*. New York: NYU Press, 2000.

Shukla, Sandya and Heidi Tinsman, eds. *Imagining Our Americas: Toward a Transnational Frame*. Durham: Duke University Press, 2007.

Silko, Leslie Marmon. *Almanac of the Dead*. New York: Simon & Schuster, Inc. 1991.

Sidbury, James. "Plausible Stories and Varnished Truths." *The William and Mary Quarterly* 59.1 (2002): 179–184.

Smith, Shawn Michelle. *American Archives: Gender, Race, and Class in Visual Culture*. Princeton: Princeton University Press, 1999.

——— . *Photography on the Color Line: W. E. B. DuBois, Race, and Visual Culture*. Durham: Duke University Press, 2004.

Smith, Theophus. *Conjuring Culture: Biblical Formations of Black America*. New York: Oxford University Press, 1994.

Snipp, C. Matthew. "Racial Measurement in the American Census: Past Practices and Implications for the Future." *Annual Review of Sociology* 29 (2003):563–588.

Sollors, Werner, ed. *Multilingual America: Transnationalism, Ethnicity, and the Languages of American Literature*. New York: NYU Press, 1998.

Sommer, Doris. *Foundational Fictions: The National Romances of Latin America*. Berkeley: University of California Press, 1993.

——— . *Proceed with Caution, When Engaged by Minority Writing in the Americas*. Cambridge, MA: Harvard University Press, 1999.

———"Supplying Demand: Walt Whitman and the Liberal Self," in *Reinventing the Americas: Comparative Studies of Literature of the United States and Spanish America*, edited by Bell Gale Chevigny and Gari Laguardia. 68–91. Cambridge: Cambridge University Press, 1986.

Sontag, Susan. *On Photography*. New York: Picador, 1977.

Spence, Jonathan. *The Chan's Great Continent: China in Western Minds*. New York: W. W. Norton and Cia., 1999.

Spillers, Hortense. *Comparative American Identities: Race, Sex and Nationality in the Modern Text*. New York: Routledge,1991.

Steckel, Richard H and Jerome C. Rose, eds. *The Backbone of History: Health and Nutrition in the Western Hemisphere*. Cambridge: Cambridge University Press, 2003.

Steedman, Carolyn. *Dust: The Archive and Cultural History*. New Jersey: Rutgers University Press, 2002.

Stephens, Michelle Ann. *Black Empire: The Masculine Global Imaginary of Caribbean Intellectuals in the United States, 1914–1962*. Durham: Duke University Press, 2005.

Stephens, Mitchell. *The Rise of the Image and the Fall of the Word*. New York: Oxford University Press, 1998.

Stigter, Bianca. "Mirrored Images: The World Reflected in Photographed Eyes." In *Questioning History: Imagining the Past in Contemporary Art*, edited by Frank van der Stok, Frits Gierstbergy and Flip Bool. Rotterdam: NAi Publishers, 2008.

Stinchcomb, Dawn F. *The Development of Literary Blackness in the Dominican Republic*. Gainesville: University Press of Florida, 2004.

Stoler, Ann Laura. *Race and the Education of Desire: Foucault's History of Sexuality and the Colonial Order of Things*. Durham: Duke University Press, 1995.

Stone, Paul. "J. Frank Dobie and the American Folklore Society." In *Corners of Texas*, edited by Francis Edward Abernathy. Dallas: University of North Texas Press, 1993. 47–66.

Storey, Rebecca. *Life and Death in the Ancient City of Teotihuacán: A Modern Paleodemographic Synthesis*. Tuscaloosa: University of Alabama Press, 1992.

Storey, Rebecca, Lourdes Márquez Morfin, and Vernon Smith. "Social Studies and the Maya Civilization of Mesoamerica: A Study of Health and Economy of the Last Thousand Years" in *The Backbone of History*, edited by Richard H. Steckel and Jerome C. Rose. Cambridge: Cambridge University Press, 2003.

Swan, Susan L. "Mexico in the Little Ice Age." *Journal of Interdisciplinary History* 11.4 (Spring 1981): 633–646.

Telles, Edward E. *Race in Another America: The Significance of Skin Color in Brazil*. Princeton, NJ: Princeton University Press, 2004.

Text Encoding Initiative. Institute for Advanced Technology in the Humanities. University of Virginia. <http://www.tei-c.org>

Therrell, Matthew D., David W. Stahle and Rodolfo Acuña Soto. "Aztec Drought and the 'Curse of One Rabbit.'" *American Meteorological Society*. (September 2004): 1263–1272.

———. David W. Stahle, José Villanueva Díaz, Eladio H. Cornejo Oviedo, and Malcolm K. Cleaveland. "Tree Ring Reconstructed Maize Yield in Central Mexico: 1474–2001." *Climatic Change* 74 (2006): 493–504.

Tillis, Antonio. "Awakening the Caribbean African: The Socio-Political Poetics of Blas Jimenez." *Afro-Hispanic Review* 2 (2003): 29–38.

Tinnemeyer, Andrea. "Enlightenment Ideology and the Crisis of Whiteness in *Francis Berrian* and *Caballero*." *Western American Literature* 35:1 (2000): 21–32.

Trachtenberg, Alan. *Reading American Photographs: Images as History Mathew Brady to Walker Evans*. New York: Hill and Wang, 1989.

Trouillot, Michel-Rolph. *Silencing the Past: Power and the Production of History*. Boston: Beacon Press, 1995.

Truett, Samuel. "Epics of Greater America: Herbert Eugene Bolton's Quest for a Transnational American History." In *Interpreting Spanish Colonialism: Empire,*

Nations, and Legends, edited by Christopher Schmidt-Nowara and John M. Nieto-Phllips, 213–247. Albuquerque: University of New Mexico Press, 2005.

Turner, Bryan S. *The Body and Society: Explorations in Social Theory.* New York: Basil Blackwell, Inc., 1984.

———. "Contemporary Problems in the Theory of Citizenship." In *Citizenship and Social Theory*, ed. B. S. Turner. 1–19. London: Sage, 1993.

Turner, Frederick Jackson. "The Significance of the Frontier in American History." In *Where Cultures Meet: Frontiers in Latin American History*, edited by David J. Weber and Jane M. Rausch, 1–19. Wilmington: Scholarly Resources Inc., 1994.

U.S. Census Bureau. *Table 1. United States—Race and Hispanic Origin: 1790 to 1990.* Washington, D.C.: U.S. Census Bureau, 2002. http://www.census.gov/population/www/documentation/twps0056/tab01.pdf (accessed on 11 January 2010).

———. *2009 National Population Projections (Supplemental).* Washington, D.C.: U.S. Census Bureau, 2009. http://www.census.gov/population/www/projections/2009cnmsSumTabs.html (accessed on 11 January 2010).

Valdez, Guadalupe, Jr. "Memories of My Grandfather: Luis G. Gómez." In *Crossing the Rio Grande: An Immigrant's Life in the 1880s.* College Station: Texas A&M University Press, 2007.

Varese, Stefano (coordinator). *Pueblos indios, soberanía y globalismo.* Biblioteca Abya-Yala, Quito, 1996.

Vázquez, Oscar E. "'A Better Place to Live': Government Agency Photography and the Transformation of the Puerto Rican Jíbaro." In *Colonialist Photography: Imag(in)ing Race and Place*, edited by Hight and Sampson, 281–310. New York: Routledge, 2002.

Velez, William and Rogelio Saenz. "Toward a Comprehensive Model of the School Leaving Process among Latinos." *School Psychology Quarterly* 16. 4 (2001): 445–467.

Venuti, Lawrence. *The Translator's Invisibility: A History of Translation.* New York: Routledge, 1995.

Verhovek, Sam Howe. "Mother Scolded by Judge for Speaking Spanish." *New York Times* 29 Aug. 1995. http://www.nytimes.com/1995/08/30/us/mother-scolded-by-judge-for-speaking-in-spanish.html (accessed on 11 January 2010).

Visweswaran, Kamala. *Fictions of Feminist Ethnography.* Minneapolis: University of Minnesota Press, 1994.

Von Eschen, Penny. *Race Against Empire: Black Americans and Anti-Colonialism.* Ithaca: Cornell University Press, 1997.

———. *Satchmo Blows Up the World: Jazz Ambassadors Play the Cold War.* Cambridge, MA: Harvard University Press, 2005.

Wallerstein, Immanuel. "The Geoculture of Development, or the Transformation of Our Geoculture." In *After Liberalism.* New York: The New Press, 1995.

———. "The French Revolution as a World-Historical Event." In *Unthinking the Social Sciences: The Limits of Nineteenth-Century Paradigms.* Cambridge: Polity Press, 1991.

Wallerstein, Immanuel. *Geopolitics and Geoculture: Essays on the Changing World-System*. Cambridge: Cambridge University Press, 1991a.

———. *The Modern World-System III: The Second Era of Great Expansion of the Capitalist World-Economy, 1730–1840s*. London: Academic Press, 1989.

———. "World-Systems Analysis." In A. Giddens and J. H. Turner eds. *Social Theory Today*. Cambridge: Polity Press, 1987.

———. *The Modern World-System II: Mercantilism and the Consolidation of the European World-Economy, 1600–1750*. London: Academic Press, 1980.

———. *The Modern World-System: Capitalist Agriculture and the Origins of the European World-Economy in the Sixteenth Century*. New York: Academic Press, 1974.

Warrior, Robert Allen. "A Room of One's Own at the ASA: An Indigenous Provocation." *American Quarterly* 55.4 (December 2003): 681–687.

Weber, David J. *The Mexican Frontier, 1821–1846: American Southwest under Mexico*. Albuquerque: University of New Mexico Press, 1982.

———. *Myth and History of the Hispanic Southwest*. Albuquerque: University of New Mexico Press, 1988.

———. *The Spanish Frontier in North America*. New Haven: Yale University Press, 1992.

Whitaker, Arthur P. "The Americas in the Atlantic Triangle." In *Do the Americas Have a Common History?, A Critique of the Bolton Theory*, edited by Lewis Hanke. 141–164. New York: Knopf, 1964.

———. *The Western Hemisphere Idea: Its Rise and Decline*. Ithaca: Cornell University Press, 1954.

Whitman, Walt. "The Spanish Element in Our Nationality." In *Poetry and Prose*. New York: Library of America, 1996.

Whittington, Stephen and David Reed, eds. *Bones of the Mayas, Studies of Ancient Skeletons*. Washington, D.C.: Smithsonian Institute Press, 1997.

Williams, Carol J. *Framing the West: Race, Gender, and the Photographic Frontier in the Pacific Northwest*. New York: Oxford University Press, 2003.

Williams, Simon J. and Gillian Bendelow. *The Lived Body: Sociological Themes, Embodied Issues*, New York: Routledge, 1998.

Williams, William Proctor and Craig S. Abbott. *An Introduction to Bibliographic and Textual Studies*. 4th ed. New York: The Modern Language Association of America, 2009.

Wolf, Eric C. *Europe and the People Without History*. Berkeley: University of California Press, 1982.

Worcester, Donald E. "The Significance of the Spanish Borderlands to the United States" in *New Spain's Far Northern Frontier: Essays in Spain in the American West, 1540–1821*. Ed. David J. Weber. Albuquerque: University of New Mexico Press, 1979. 1–14.

Wu, Frank H. *Yellow: Race in American Beyond Black and White*. New York: Basic Books, 2002.

Zamora, Lois Parkinson. *The Usable Past: The Imagination of History in Recent Fiction of the Americas*. Cambridge: Cambridge University Press, 1997.

Zamora, Lois Parkinson. *Writing the Apocalypse: Historical Vision in Contemporary U.S. and Latin America Fiction*. Cambridge: Cambridge University Press, 1993.

Zavala, Silvio. "The Frontiers of Hispanic America" in *Where Cultures Meet: Frontiers in Latin American History*. Eds. David J. Weber and Jane M. Rausch. Wilmington: Scholarly Resources Inc., 1994. 42–50.

Zea, Leopoldo. "The Interpenetration of the Ibero-American and North American Cultures." *Philosophy and Phenomenological Research* 9.3 (1949): 538–544.

———. *The Role of the Americas in History*. Trans. Sonja Karsen. Savage: Rowman & Littlefield, 1992.

Zimmerman, Arthur Franklin. *Francisco de Toledo: Fifth Viceroy of Peru, 1569–1581*. Caldwell: The Caxton Printers, Ltd., 1938.

Author Biographies

Melissa Bailar received her PhD in French studies from Rice University. She is currently associate director of the Humanities Research Center at Rice and also works as a researcher on the Our Americas Archive Partnership. She serves on the executive board of the South Central Modern Language Association and regularly writes and presents on nineteenth- and twentieth-century French and Francophone literature and film.

Antonio Barrenechea is assistant professor of English at the University of Mary Washington. His recent work includes "Hemispheric Horrors: Celluloid Vampires from the 'Good Neighbor' Era," published in *Comparative American Studies*. He is currently completing "Hemispheric Legacies: New World Archives in Encyclopedic Novels," a book-length manuscript that examines the incorporation of colonial histories in North American fiction.

Deborah Cohn is associate professor of Spanish at Indiana University—Bloomington. She co-edited *Look Away!: The U.S. South in New World Studies* with Jon Smith (Duke, 2004) and is author of *History and Memory in the Two Souths: Recent Southern and Spanish American Fiction* (Vanderbilt, 1999). She is currently working on a book titled *Creating the Boom's Reputation: The Promotion of Latin American Literature in and by the U.S. during the Cold War*.

Michael O. Emerson is the Allyn and Gladys Cline Professor of Sociology and co-director of the Institute for Urban Research at Rice University. He is the author of several books, including the award-winning *Divided by Faith: Evangelical Religion and the Problem of Race in America* (Oxford, 2000) and *People of the Dream: Multiracial Congregations in the United States* (Princeton, 2006). His most recent books are *Transcending Racial Barriers: Toward a Mutual Obligations Approach* (Oxford, 2010) and *Religion Matters: What Sociology Teaches Us about Religion in Our World* (Allyn & Bacon, 2010).

Matthew Pratt Guterl is professor of African American and African Diaspora studies, and director of American studies at Indiana University. He is the author of *The Color of Race in America, 1900–1940* (Harvard, 2001) and *American Mediterranean: Southern Slaveholders in the Age of Emancipation* (Harvard, 2008), and the co-editor (with James T. Campbell and Robert G. Lee) of *Race, Nation, and Empire in the Age of Emancipation* (UNC, 2007).

Ruth Hill is a professor of Spanish at the University of Virginia, where she teaches courses in the Spanish and English departments. The author of two books and numerous essays, she is currently working on the relations between folk biology, natural history, and human diversity in the early modern Hispanic world, and on a history of Aryanism in the Americas.

Rodrigo Lazo is associate professor of English at the University of California Irvine. He is the author of *Writing to Cuba: Filibustering and Cuban Exiles in the United States* (UNCP, 2005), and he is currently working on a book-length study of the hemispheric dimensions of Spanish-language print culture in the early nineteenth-century United States.

Caroline F. Levander is the Carlson Professor in the Humanities, professor of English, and director of the Humanities Research Center at Rice University. She is author of *Cradle of Liberty: Race, the Child, and National Belonging from Thomas Jefferson to W. E. B. Du Bois* (Duke, 2006), *Voices of the Nation: Women and Public Speech in Nineteenth-Century American Literature and Culture* (Cambridge, 1995, 2009), as well as co-editor of *Hemispheric American Studies* (Rutgers, 2008). She is co-editing, with Walter Mignolo, a special issue of *The Global South: World Dis/Order* (2011).

Moramay López-Alonso is an assistant professor in the Department of History at Rice University. Although she is an economic historian, whose research has focused on Mexico's biological standards of living, Moramay's interests generally encompass Latin American culture, literature, and politics (as well as history).

Walter D. Mignolo is the William H. Wannamaker Professor of Literature and Romance Studies, professor of cultural anthropology and professor of Spanish and Latin American studies, and director of the Center for Global Studies at Duke University. He is the author of *Local Histories/Global Designs: Coloniality, Subaltern Knowledges and*

Border Thinking (Princeton University Press, 2000), *The Darker Side of the Renaissance: Literacy, Territoriality and Colonization* (Michigan, 1995), and *The Idea of Latin America* (Blackwell, 1991), among other books and edited collections. He is co-editing with Caroline Levander a special issue on the "Global South: World Dis/Order."

Heather Miner is a doctoral candidate in the Department of English at Rice University, as well as a graduate certificate candidate in the Center for the Study of Women, Gender and Sexuality. She has published reviews and articles on narrative, domesticity, and nationalism. Her current project studies Victorian regionalism and mid-nineteenth-century reform politics in Britain. He is co-editing with caroline Levander a special issue on the "Global South: World Dis/Order".

Anthony B. Pinn is Agnes Cullen Arnold Professor of Humanities and professor of religious studies at Rice University. He is also director of research for the Institute of Humanist Studies Think Tank. His interests include religion and popular culture, constructive theologies, religious aesthetics, and African American humanism. He is the author/editor of twenty-three books, including *Embodiment and the New Shape of Black Theological Thought* (NYUP, 2010), *Black Religion and Aesthetics* (Palgrave, 2009), and *Terror and Triumph: The Nature of Black Religion* (Fortress, 2003).

Robin Sager is a PhD candidate in history at Rice University. She is currently completing her dissertation, which is a comparative study of marital cruelty in antebellum Virginia, Texas, and Wisconsin.

Karen Manges Douglas is an associate professor in the Department of Sociology at Sam Houston State University. Her research focuses on racial and ethnic relations, social inequality, welfare reform, and water resources and the environment.

Rogelio Saenz is a professor in the Department of Sociology at Texas A&M University. His research focuses on demography, immigration, social inequality, and racial and ethnic relations. He is co-editor of the book *Latinas/os in the United States: Changing the Face of América* (Springer, 2007).

Index

"Black Atlantic," 34
"Carta de Jamaica" (Bolívar), 57
A Chronicle of Old Florida and Southwest (Bolton), 15
"Defensive Spanish Expansion and the Significance of the Borderlands" (Bolton), 15
"The Epic of Greater America" (Bolton), 13–16, 29
"Have We a National Literature" (Whitman), 37
"Hegel y el moderno panamericanismo" (O'Gorman), 20
"The Herald" (Hawthorne), 37
"Los Vecinos/The Neighbors" exhibit, 99
"nation," 13–20
"North and South America" (Sarmiento), 129
A Plague of Sheep (E. Melville), 168
"Rappaccini's Daughter" (Hawthorne), 38
"The Significance of the Spanish Borderlands to the United States" (Worcester), 18
A Texas-Mexican Cancionero: Folksongs of the Lower Border (Paredes), 18
"Voyage dans la République de Colombia" (Mollien), 234

Abu-Lughod, Janet L., 72
Academic Culture in the Spanish Colonies (Lanning), 18
Adams, John, 31
Adams, Rachel, 41, 129
African Diaspora, African Diaspora Studies, 3, 43–45, 60, 104, 247, 250, 255
African slaves, 50, 54–55, 64

African-American Studies, African American history, 28, 132, 217
Agassiz, Louis, 96–97
Aiton, Arthur Scott, *Antonio de Mendoza: First Viceroy of New Spain*, 18, 263
Alamo, 128–129, 237
Alger, Horatio, 130, 256
Algería, Fernando, 37
Almanac of the Dead (Silko), 54
Alonzo, Armando, 192
Alva, Klor, de, 61
Ambassadors of Culture (Gruesz), 42
América y sus mujere, 226
American hemisphere, 1, 3, 23, 43, 93–96, 98, 102, 104, 222, 226
American Historical Association (AHA), 13, 15, 29, 184
American Journal of Human Genetics, 127
American Literary Studies, 22–23, 38, 40
American Mediterranean: Southern Slaveholders in the Age of Emancipation, 35
American Quarterly, 260
American Studies, pedagogy, 3, 7–8, 244
American Studies Association (ASA), conference, 2, 29, 245, 252, 260
The Americas: A Hemispheric History (Frernández-Armesto), 22
Amerindian, 16, 18, 25, 55, 58–60, 64, 160–161, 165, 166, 168, 171
See also Native Americans
Ampuero, Nicolas, 234
Anderson, Robert, *Story of Extinct Civilizations of the East,* 123

Anglo-American groups, 181, 185–186, 188, 190
Anglo-Saxon, Anglo-Saxon Clubs, 36, 56, 58–59, 63, 111, 117–118, 131, 144
Annals of Human Genetics, 127
Antonio de Mendoza: First Viceroy of New Spain (Aiton), 18
Anzaldúa, Gloria, 19, 55
Anzaldúa, Gloria, 236; *Borderlands/La Frontera: The New Mestiza,* 19
The Archaeology of Knowledge (Foucault), 205
Archive Fever, 44
See also *Mal d'archive*
Archives; national archives; translation, 200, 203, 205, 213
Archivo arroz Américas, 231
See also Rice University
Arrighi, Giovanni, 52
Articles of Confederation, 31
Aryan Nations, 110, 131
Asian-American history, 132
Atlantic commercial circuit, 50–56, 64, 65, 68, 70
Austin, Stephen F., 223
Ayala, Cristina, 43
Ayala, Guaman Poma de, 61

Baldwin, James, *The Evidence of Things Not Seen; The Fire Next Time,* 130
Balfe, Michael William, *The Bohemian Girl,* 193
Barker, Eugene C., 184
Barlow, Joel, *The Vision of Columbus,* 25
Baroque Times in Old Mexico (Leonard), 18
Beinecke library, 213
Benavente, Toribio de (Motolinía), 159–160
Black Legend, 15, 18, 57, 157, 159, 161, 171
Black Scare, 111, 116, 118, 126, 129, 132
Blithedale Romance, The (Hawthorne), 38
Boarh, Woodrow, 45, 162
Boas, Franz, 112
Bobby, 129
the body, 93–95
"black bodies," 93, 96, 101
"white bodies," 99
Bohemian Girl, The (Balfe), 193
Bolívar, Simón; "Carta de Jamaica," 57

Bolton, Herbert E., "The Epic of Greater America," 13–16, 29
Bolton, Herbert Eugene, "The Epic of Greater America;" "Defensive Spanish Expansion and the Significance of the Borderlands;" *The Spanish Borderlands; A Chronicle of Old Florida and Southwest, History of the Americas: A Syllabus with Maps,* 15
border policies, 86
Border studies, 4, 15, 17–20
Borderlands/La Frontera: The New Mestiza (Anzaldúa), 19
Borges, Jorge Luis, 37, 66
Brack, Gene M, 129
Bradford, Edward, 116
Brazil, 24, 35, 43, 78, 80, 82–83, 96, 101
Brickhouse, Anna, *The Story of Don Luis de Velasco,* 42
British Atlantic, 109, 111
British East India Company, 38
Brown, Charles Brockden, 41
Brown Scare, 129–130
Buchanan, Patrick, 131
Burton, María Amparo Ruis de; *Who Would Have Thought It?; The Squatter and the Don,* 210
Business Week, 69
Bustamante, Jorge, 219, 221, 240
Buxton, Dudley L.H., *The Peoples of Asia,* 122

Caballero (González and Eimer), 179–180, 186–193
Caldas, Francisco José de, "Viaje de Quito a Popayan," 234
Calderón, Francisco, *Les Democraties Latines,* 116
Callahan, Monique-Adelle, 43
Cañizares-Esguerra, Jorge, *Puritan Conquistadors: Iberianizing the Atlantic, 1550–1700,* 22
Capac, Huayna, 166
capitalism, 131, 191
global, world system of, 51–53, 68–69
Caribbean, 24–25, 43, 49, 53, 58–60, 70, 154, 221, 247, 251
Studies, 248, 250, 254
Carlota of Mexico (empress), 237

Casis, Lilia, 181
Cassanova, Pablo Gonzaléz, 54
Casteñeda, Carlos E., 184
Caucasian, 109, 112, 118–123, 125–126, 129, 132
Cecilia Valdes (Villaverde), 42
Cervantes, Miguel de, *La Gitanilla,* 193
Charles V (Spanish king and Holy Roman Emperor), 160, 165
Cherniavsky, Eva, 245
Chevigny, Bell Gale, 30
Chicano studies, 18, 180
Christian Identity, 127–130, 131
Civil Rights Movement, 104, 152
Civil War, 30, 32, 39, 61, 128–129, 167, 187
Clashing Tides of Colour (Stoddard), 112
class
  consciousness, 35
  crossing boundaries of, 147
  identification, 181
  intersection of, 130–131
  as signifier, 129
Clinton Institute, 256
Cohn, Deborah, *Look Away!: The U.S. South in New World Studies,* 34
Cold War, 247, 250, 254, 259–260
colonialism, 51, 55, 58–59, 66–69, 79, 100, 199, 237
  biological encounter with, 158, 163
  smallpox as a result of contact, 157, 164–166, 169–170
  Spanish colonialism, 161
  and the spread of syphilis, 157, 164, 167
coloniality, 30, 51–56
  contrasted with modernity, 63–64
  of power, 65, 68
color line, 111, 112, 117, 119, 127, 139–140, 147, 151, 152
Columbia University, 182
Columbian Exchange, The (Crosby), 17, 157
Columbus, Christopher, 25, 159, 164, 172
Communication and Cultural Studies, 250
Connexions platform, 238–239
Conquest of A Continent or the Expansion of Races in America, The (Grant), 112, 114

Cook, Sherburn, 162
Cooper, James Fenimore, 42
Coronado, Francisco Vasquez de, 15
Cortés, Hernán, 25, 160, 162
Cotera, María, 179–180, 184
Cox, Earnest Sevier, *White America,* 118, 129
Cox, Renée, "It Shall Be Named," 102
Creole, 54, 56, 58–61, 64, 79, 114
  black Creole consciousness, 58–59, 61
  Creole double consciousness, 56, 60, 70
  Hispanic Creole consciousness, 58
Critical Race Theory (CRT), 44, 110
Crosby, Alfred W., 17, 157
Crosby, Alfred W., Jr., *The Columbian Exchange,* 157
Crossing the Rio Grande (Kreneck), 202–206, 208–209
  See also *Mis Memorias*
CRT. *See* Critical Race Theory
Cuba, 24–25, 35, 42–43, 47, 80, 141, 212, 239
Cuitlahuactzin, 165
Cullen, Henry, 57
Cummins, Maria, 41

Darío, Rubén, *La Oda a Roosevelt,* 63
Dartmouth University, 256
Darwin, Charles, 120
decolonization, 53, 61, 64
Delany, Martin, 41
Deloria, Phil, 249
Democracy in America (Tocqueville), 39
Democracy Studies, 245
Democraties Latines, Les (Calderón), 116
DePelchin, Kezia, 237
Derrida, Jacques; *Mal d'archive,* 205
Dessalines, Jean-Jacques, 60
Dew on the Thorn (González), 185–187, 193
Dimock, Wai Chee, 43
diversity, 21–22, 54, 93, 179, 212
  bio-, 157, 165
  cultural, 153
  racial/ethnic, 32, 87, 90–92, 152
Dixon, Roland, *Racial History of Man,* 121
Dobie, J. Frank, 182

Dorson, Richard, 246
double consciousness, 55–56, 59–61, 70
Douglass, Frederick, 41
Doyle, Don, 30
Drago, Luís María, 61–62
Du Bois, W.E.B., 278
Duke University, 245
Dunham, George, 223, 236

EADA (Early American Digital Archive), 221
Early Ibero/Anglo Americanist Summit, 221
Eimer, Margaret, 186–187, 189
El Yara (Martí), 239
Elliott, J.H.; *Empires of the Atlantic World: Britain and Spain in America*, 22
Enoch, Reginald; *Secret life of the Pacific*, 119
The Epic of America (Truslow), 13
Estevez, Emilio, 129
eugenics, 110, 112, 118, 132, 143, 181
European Union, 68
Evangeline: A Tale of Acadie (Longfellow), 25
Evidence of Things Not Seen, The (Baldwin), 130

femininity, 100
Ferdinand of Castile (Spanish king), 127
Fernández-Armesto, Felipe; *The Americas: A Hemispheric History*, 22
Fire Next Time, The (Baldwin), 130
Firmat, Gustavo Pérez, 30
Fitz, Earl, 23
Forbes, Jack, 129
Forest, Lacandon, 54
Foucault, Michel
 The Archaeology of Knowledge, 205
 The Order of Things, 66
Fox, Claire, 23
Francis, Samuel, 131
Francisco de Toledo: Fifth Viceroy of Peru (Zimmerman), 18
French Revolution, 51, 63–65
Freud, Sigmund, 205–206
Freud's Moses: Judaism Terminable and Interminable (Yerushalmi's), 205
Fuller, Margaret, 37

Galinsky, Hans, 30
Galvenston, TX, 39
Garza-Falcón, Leticia Magda, 180

Gauthereau-Bryson, Lorena, 232, 234, 238
gender studies, 28, 44
geo-culture, 51, 61, 63–65
Gerbi, Antonello, 30
Gieryn, Thomas, 245
Giles, Paul, 28
Gilroy, Paul, 34, 70
Gish, Lillian, 190
Glissant, Eduardo, 49–50
Goldberg, Michael, 249
Gómez, Luis G.; *Mis Memorias,* 201, 206
González, John, 192
González, Jovita
 Caballero, 179–180, 186–193
 Dew on the Thorn, 185–187, 193
 Social Life in Webb, 183
 Starr and Zapata Counties, 183
Grant, Madison
 The Conquest of A Continent or the Expansion of Races in America, 112
 The Passing of the Great Race, 112
Great Depression, 82, 114
Gruesz, Kirsten Silva; *Ambassadors of Culture*, 42
Guevara, Ernesto "Che," *Motorcycle Diaries*, 247
Guterl, Matthew
 Look Away!: The U.S. South in New World Studies American Mediterranean, 34
Gutiérrez, Ramón A., 211

Hale, Edward Everett, 40
Hall, Stuart, 111
Handley, George, 42–43
Hanke, Lewis, *Do the Americas Have a Common History?*, 20, 30
Harper, Frances, Ellen Watkins, 43
Harvard University
 Louis Agassiz, 96–97
Harvey, Bruce, U.S. National Narratives and the Representation of the Non-European World, 1830–1865, 41
Hatian Revolution, 51, 58, 60, 64
Hawthorne, Julian, "The Future of America," 40

Hawthorne, Nathaniel
  "The Herald," 37
  The House of the Seven Gables, 38
  Life of Franklin Pierce, 38
  "Rappaccini's Daughter," 38
  The Scarlet Letter, 38
Hemispheric American Studies (Levander and Levine), 3, 22, 28, 209, 243, 259
hemispheric studies
  multidisciplinarity, 23
  teaching approaches, 2, 5–7, 28, 41, 44
Higginson, Thomas Wentworth, "Americanism in Literature," 32
Historia moral de las mujeres, 226
History of the Americas: A Syllabus with Maps (Bolton), 17
History of the Conquest of Mexico (Prescott), 17
History of the Conquest of Peru (Prescott), 18
Hitler, Adolf, 112
House of the Seven Gables, The (Hawthorne), 38
Hrdlička, Aleš, *Remains of Eastern Asia of the Race That Peopled America,* 123–125
Human Biology, 127
Human Immunology, 127
Humboldt, Alexander von, 57, 119
Huntington, Samuel, 66

Iberian Atlantic, 109, 111
IMLS. *See* Institute of Museum and Library Services
An Immigrant's Life in the 1880s, 209
Immigration
  black immigration, 114
  Mexican immigration, 114
  West Indian immigration, 130
Immigration Reform and Control Act, 86
imperialism, 20, 116, 129
Inca Empire, 167
Indian. *See* Native Americans
Indiana University (IU): Center for Latin American and Caribbean Studies
  Ethnic Studies, 257–258
  graduate studies, 253
  International Studies, 251, 254, 255, 259

Native American and Indigenous Studies (NAIS), 258
Responsibility-Centered Management (RCM), 258
indigenismo. See racial myth
indigenous peoples, 257
Indología (Vasconcelos), 118
Institute of Museum and Library Services (IMLS), 222
Instituto Mora, 221–222, 226
integration, 17, 61, 83, 126–128, 130, 132, 148, 187, 191–193, 209–210, 238
interdisciplinary studies, 244–246, 248, 250–252, 254, 258
internal colonialism, 59
International Council of Archives, 199
Into the Darkness: Nazi Germany Today, 112
The Invention of America: An Inquiry into the Historical Nature of the New World and the Meaning of Its History (O'Gorman), 20–21
Isabella of Castile (Spanish queen), 127

Jacobson, Matthew Frye, 130
Jamestown, Virginia, 42, 187
Jefferson, Thomas, 25
Jesuits, 16, 50, 68
Jicoténcal, 210
Jim Crow, 126
Johnson-Reed Act, 118
Judaism, 205–206

Kennewick Man, 118
Kerouac, Jack, *On the Road,* 247
Khmers, 121
Kissinger, Henry, 69
Kreneck, Thomas H., 202–203
Ku Klux Klan, 131
Kuhn, Annette, 98
Kutzinski, Vera, 30

*La hija de Rappaccini (*Paz), 38
La Plata, 14
labor studies, 203
Laguardia, Gari, 30
Lanning, John Tate; *Academic Culture in the Spanish Colonies,* 18

Lapouge, Vacher de, 112
Las Casas, Bartolomé de, 159
Las Casas, Bartolomé de, *The Short Account*, 160, 171
Las Mujeres españoles, portuguesas, y americanas, 225–226, 234, 236, 238
Latin America, 4–5, 15, 17–18, 20–22, 24, 129, 132, 139–140, 155, 191, 213, 221, 231, 246–248, 250–251, 254, 255, 260
Latin American Studies
  Latin American and Caribbean Studies, 248, 250, 254–255
  Latino Studies, 19, 118, 255
Latinophobia, 115, 129, 130, 132
Lazo, Rodrigo, 5–6, 42, 199
LCSH. *See* Library of Congress
Ledoux, Cory, 238–239
Leonard, Irving A., *Baroque Times in Old Mexico*, 18
Les Democraties Latines (Calderón), 116
Levander, Caroline F., *Hemispheric American Studies*, 3, 22, 28
Levine, Robert S., *Hemispheric American Studies*, 3, 22, 28
Library Company of Philadelphia, 213
Library of Congress, subject headings (LCSH), 237
Life of Franklin Pierce (Hawthorne), 38
Limón, José, 179
Lincoln, Abraham (American president), 129
Lomas, Laura, 42
Lone Star (Adams), 129
Longfellow, Henry Wadsworth, 241; *Evangeline: A Tale of Acadie*, 25
Look Away!: The U.S. South in New World Studies (Cohn and Guterl), 34
Louisiana, 34, 181
l'Ouverture, Toussaint, 60

Madison, James, 31
Mahori, 121
Mal d'archive (Derrida). See also *Archive Fever*, 205, 207, 216, 267
Manoff, Marlene, 203
March of the Titans (Stormfront), 127, 137
Martí, José: *El Yara;* "Our America," 239

Maryland Institute for Technology in the Humanities (MITH), 221
masculinity, 100
Massey, Doug, 80, 83, 154–155
Mauad, Ana Maria, 101
Maximilian of Mexico (emperor), 237
McAllister, Kirsten Emiko, 98
McCaa, Robert, 171
McClennen, Sophia, 27
MeCha, 127
melting pot, 111–112, 114, 116
Melville, Elinor
  A Plague of Sheep, 168
  Valle del Mezquital, 168
Melville, Herman, 41–42, 214
Memorias de mi Viaje (Recollections of My Trip) (Torres), 39
memory, 59, 70–71, 79, 199, 201–205, 209, 211
  historical/cultural, 103
  national, 209
Menchú, Rigoberta, 54–55
Mesoamerica, 5–7
  Tenochtitlán, 7
  Teotihuacán, 7
  See also Native Americans
mestizaje, 19, 60, 79, 109, 111–112, 114, 118
Mexican American folklore, 180
Mexican American integration, 83, 210
  cultural, 191, 193
  of language, 209
  in Texas, 187, 191, 192
  and white supremacist ideology, 112, 126–128, 130, 132
Mexican American War, 128–129, 152, 190
Mexican Migration Project, 84
Mexico
  Conquest of Mexico, 17, 160, 164
  Mexican Revolution, 39, 118
  Republic of Mexico, 39
Mexico City (Tenochtitlán), 6, 19, 162–163
Mignolo, Walter, 4, 30–31
migrant archives, 200–201, 204, 206, 208–209, 212–214

migration, 38, 41, 78, 80–81, 82, 83, 84, 85, 90, 201
   Aztec, 19
   English, 58
   forced, 171
   Latin American, 68–70
   racial, 28
Mireles, Edmundo E., 186
Mis Memorias (Gómez), 201–203, 205, 208–210
   See also *Crossing the Rio Grande*
miscegenation, 25, 115, 117, 126–129
Mitchell, Margaret, *Gone With the Wind*, 190
MITH. *See* Maryland Institute for Technology in the Humanities
modernity, 4, 51–55, 63–67, 95, 100
Mollien, Gaspard-Thoodore, "Voyage dans la République de Colombia," 234
Mongolian, 234, 238
Monroe Doctrine, 27, 49, 62–63
Montejano, David, 181, 192
Moretti, Franco, 26
Morgan, Jennifer, 95
Morieras, Alberto, 28
Morrison, Toni, *Playing in the Dark*, 130
Motolinía. *See* Benavente, Toribio de
Motorcycle Diaries, (Ridley) (Walter), 247
Mujeres espanolas, portuguesas, y americana, Las, 226, 234, 238
mulatto, 4, 17, 21, 111, 122, 126, 140
The Multilingual Anthology of American Literature (Shell and Sollors), 204
multinationalism, 1–4, 8, 21, 23–24, 28–29

Nahuatl, 160, 165–166, 169
nation, 2–3, 5–9, 13–19, 23–25
   national identity, 31, 32, 38, 39, 100
   national narrative, 3
National Endowment for the Humanities (NEH), 26, 241, 274
national identity. *See* nation
National narrative. *See* nation
Nationalism, 22, 30, 31, 35, 36, 38, 40, 41, 44, 158, 193, 237
   national borders, 1–2, 4, 7–8, 21–23
Native American Studies, 16–18, 44

Native Americans, 1, 3–8, 14, 16
   Population, 1, 3–17, 45, 158, 159–164
   role of smallpox, 8, 15, 164–166, 169–170
nativists movements. *See* white supremacy
Nazism, 66
Nebrija, Antonio de, 160
Negrophobia, 6–7, 21, 24, 26, 30
NEH (National Endowment for the Humanities), 3
Neruda, Pablo, 37, 43
New York Times, 116, 145
North Atlantic, 50, 53, 69–71, 170

O'Gorman, Edmundo
   "Hegel y el moderno panamericanismo," 20
   The Invention of America: An Inquiry into the Historical Nature of the New World and the Meaning of Its History, 20
Ohio State University, 245
On the Road, (Kerouac), 247
OOAP. *See* Our Americas Archive Partnership
otherness, 219, 223
Our Americas Archive Partnership (OAAP), 3–8, 28–29
   digitization process, 13–16, 21–26
   search terms, 221–223, 229–240
   Transcription, 9, 13
   translation process, 15–21, 24
Our Lady of the Lake College, 182

Padilla, Genaro M., 211
Pamplona, Marco Antonio, 30
Panama Canal, 61, 181
Paredes, Américo, *A Texas-Mexican Cancionero: Folksongs of the Lower Border*, 18–19
Paris Exposition of 1900, 100
Passing of the Great Race or The Racial Basis of European History, The (Grant), 112, 116
Paz, Octavio, *La hija de Rappaccini*, 38
Pendergraft, Rachel, 129
Peoples of Asia, The (Buxton), 122
Pérez, Vincent, 191
Peru, role of smallpox, 166
Philip II (Spanish king), 160
photography, 96, 98–102, 104

Pick, Daniel, 116
Pierce, Franklin, "Journal of his March from Vera Cruz," 38
Pizarro, Francisco, 166–167
Playing in the Dark (Morrison), 130
pluralism, 30
Poincaré, Raymond (president), 116
Poma, Guaman, 55, 61
Ponce de León, Juan, 15
postcolonial studies, 34, 59, 62, 231, 257
Powell, John, 118
Pradt, Abe de, 57
Prescott, William H., *History of the Conquest of Mexico; History of the Conquest of Peru*, 17
psychoanalysis, 205–206
Puerto Rico, 87, 141
Puritan Conquistadors: Iberianizing the Atlantic, 1550–1700, (Cañizares-Esguerra), 22
Puritan history, 38

Quijano, Aníbal, 51–52

Race, 139–144, 147–149, 151–152, 154–155, 190
 racial nominalism, 122, 132
 racial purity, 109, 136
 scientific racism, 6, 16
Racial History of Man (Dixon), 121
racial hybridity, 30, 109, 117, 118, 132
Racial Integrity Law of Virginia, 118
racial migration, 28
racial myth, 1–2, 8, 16, 19, 24–25
 indigenismo, 1, 110, 118
 mestizaj, 1, 5–6
Radway, Janice, 244–245
Raza cósmica (Vasconcelos), 118
Recovering the U.S. Hispanic Literary Heritage project, 201, 210
religious studies, 3, 4, 6, 91, 250, 259
Remains in Eastern Asia of the Race That Peopled America (Hrdlička), 123–125
Rice Americas Archive. *See* Our Americas Archive Partnership, Rice University
Rice University: Rice Americas Archive (OAAP), 3–9
 Archivo arroz Américas, 16

Rising Tide of Color Against White World-Supremacy (Stoddard), 112–113
Robinson, 129
Rockerfeller Foundation, 185–186
Rodó, Enrique, "Ariel," 63
Roediger, David R., *Wages of Whiteness*, 130
Roma, Texas, 181
Roosevelt, Franklin, 20
Roosevelt, Theodore; "corollary," 61–63, 67, 69

Salles, Walter, *Motorcycle Diaries*, 247
San Diego Museum of Photographic Arts, 99
Santa Anna, Antonio López de (Mexican president), 39
Sarmiento, Domingo F. (president), 25
 "North and South America," 25, 129
Scarborough, Dorothy; *The Wind*, 190
Scarlet Letter, The (Hawthorne), 38
Scott, Rebecca, 34
Scott, Ridley, *Thelma and Louise*, 247
Secret Life of the Pacific (Enoch), 119
segregation, 4, 149
Sepúlveda, Juan Ginés de, 50–51, 55
Shank, Barry, 245
Shell, Marc,*The Multilingual Anthology of American Literature*, 204
Short Account, The (Las Casas), 160, 171
Shukla, Sandya, *Imagining the Americas*, 3, 85
Silko, Leslie Marmon, *Almanac of the Dead*, 54
slavery, 2, 9, 21, 22–23, 25, 28
 slave trade, 34, 82, 95
Smith, John, 21
Smith, Jon, 34
Smith, Shawn Michelle, 100
Smith, Theophus, 102
Smithsonian Museum, 123–125
Sollors, Werner, *The Multilingual Anthology of American Literature*, 41, 204
Sommer, Doris, 37, 191
Sontag, Susan, 99, 104
South Carolina, 96
Southwest Review, 182
Southwest United States, 114, 128–129, 182
Southwestern Historical Quarterly, 184

Souza, Auta de, 43
Spanish borderlands, 15, 18–19
The Spanish Borderlands (Bolton), 15
The Spanish Frontier in North America (Weber), 19
Spanish-American War, 18, 61, 63, 67
The Squatter and the Don (Burton), 210
Squier, Ephraim G., 41
Steedman, Carolyn, 207–208
Stephens, John L., 41
Stevens, David H., 185
Stoddard, Lothrop, *Rising Tide of Color Against White World-Supremacy*, 112
Stoler, Ann, 95
Stormfront, 29
   March of the Titans, 2
   Stormfront White Nationalist Community website, 22
Story of Don Luis de Velasco, The (Brickhouse), 42
Story of Extinct Civilizations of the East (Anderson), 123
Stowe, Harriet Beecher, 42
Strong, George Templeton, 32
Subbaswamy, Kumble, 245
Sumner, Charles, "Are We a Nation," 32

Taylor, Jared, 29
TEI. *See* Text Incoding Initiative
Tepoztlán Institute, Mexico, 8, 256
Texas, 5, 77, 78, 86, 145
   border/Border studies, 18, 39, 179, 181
   Republic of Texas, 39, 81
   Revolution, 190
   South, 155
Texas A&M University, Corpus Christi, 202
Texas Coahuila borderlands, 39
Texas Folklore Society (TFS), 182–184, 186
Texas history, 190, 203
Text Incoding Initiative (TEI), 13
TFS. *See* Texas Folklore Society
Thelma and Louise, (Scott), 247
Tinsman, Heidi, *Imagining the Americas*, 3, 85
Tocqueville, Alexis de, *Democracy In America*, 39
Torres, Olga Beatriz, *Memorias de mi Viaje (Recollections of My Trip)*, 39

transnationalism, 78, 81, 246, 259
Tressler, 55
Trouillot, Michel-Rolph, *Silencing the Past: Power and the Production of History*, 58
Truslow, James, *The Epic of America*, 13
Turner, Frederick Jackson, 15

U.S. Department of Education, 248, 251
U.S. Hispanic literary heritage, 194, 201, 210, 213
U.S.-Mexican borderlands, 179
U.S. National Archives, 205
   US National Archive Web page, 200
U.S. South, 34, 35
Uncle Tom's Cabin (Stowe), 223, 239
University of California, Berkely, 17, 162
University of Guyana, Georgetown, 250
University of Houston, 210
University of Texas, Austin (UT), 181

Valadares (govenor), 78
Valdez, Guadalupe, 202
Valle del Mezquital (E. Melville), 168–169
Varnum, James M., 31
Vasconcelos, José, *Indología*; *Raza cósmica*, 118
Vega, Garcilaso de la, 61
Velazquez, Loretta Janeta, *The Woman in Battle*, 39
Venezuela, 61–62, 83
Venuti, Lawrence, 208
Viaje a lost Estados Unidos del Norte América (Journey to the United States) (Zavala), 39–40
Villaverde, Cirilo, *Cecilia Valdes*, 42
Vine, Deloria, Jr., 70
The Vision of Columbus (Barlow), 25

Wages of Whitness (Roediger), 130
Walcott, Derek, 43
Wallace, Alfred Russell, 120
Wallerstein, Immanuel, 51–52, 63–65, 68–69
Warrener's, 191, 193
Weber, David J.; *The Spanish Frontier in North America*, 19
West Indies, 37, 59, 114

Western Hemisphere, 3–4, 7, 13, 15–16, 30, 34, 43, 49–50, 52–53, 56, 58–68, 69–71, 157–159, 162, 167–168, 170–172
Wheatley, Phillis, 237
Whitaker, Arthur P., 21
White America (Cox), 118–124, 125–129
White Aztlan, 110, 118, 126, 128
white supremacy, 4, 7, 99, 101, 118, 126, 130–131
　nativist movements, 112
Whitman, Walt, "Have We a National Literature," 37
Who Would Have Thought It? (Burton), 210
Williams, Carol, 97
Williams, Wayne, 130
Williams, William, 31
Woman in Battle, The (Velazquez), 39

Worcester, Donald E., "The Significance of the Spanish Borderlands to the United States," 18
Wu, Frank: *Yellow,* 115

Yellow (Wu), 115, 146–147, 155–156
Yellow Peril, 111, 115–116, 118, 122, 126, 132
Yenisei Ostiaks, 125
Yerushalmi, Yosef Hayim; *Freud's Moses,* 205
Yucatán peninsula, 163

Zamora, Lois Parkinson, 26, 30
Zavala, Lorenzo de, *Viaje a lost Estados Unidos del Norte América (Journey to the United States),* 39
Zea Leopoldo, 22
Zimmerman, Arthur Franklin, *Francisco de Toledo: Fifth Viceroy of Peru,* 18

GPSR Compliance
The European Union's (EU) General Product Safety Regulation (GPSR) is a set of rules that requires consumer products to be safe and our obligations to ensure this.

If you have any concerns about our products, you can contact us on

ProductSafety@springernature.com

In case Publisher is established outside the EU, the EU authorized representative is:

Springer Nature Customer Service Center GmbH
Europaplatz 3
69115 Heidelberg, Germany

www.ingramcontent.com/pod-product-compliance
Lightning Source LLC
LaVergne TN
LVHW051914060526
838200LV00004B/147